STUDENT'S
SOLUTIONS MANUAL

STATISTICS
FOR THE LIFE SCIENCES
FIFTH EDITION

Myra Samuels
Purdue University

Jeffrey A. Witmer
Oberlin College

Andrew Schaffner
California Polytechnic State University

Boston Columbus Hoboken Indianapolis New York San Francisco
Amsterdam Cape Town Dubai London Madrid Milan Munich Paris Montreal Toronto
Delhi Mexico City São Paulo Sydney Hong Kong Seoul Singapore Taipei Tokyo

The author and publisher of this book have used their best efforts in preparing this book. These efforts include the development, research, and testing of the theories and programs to determine their effectiveness. The author and publisher make no warranty of any kind, expressed or implied, with regard to these programs or the documentation contained in this book. The author and publisher shall not be liable in any event for incidental or consequential damages in connection with, or arising out of, the furnishing, performance, or use of these programs.

Reproduced by Pearson from electronic files supplied by the author.

ISBN-13: 978-0-321-98969-7
ISBN-10: 0-321-98969-4

1 2 3 4 5 6 OPM 18 17 16 15 14

www.pearsonhighered.com

PEARSON

Contents

Chapter 1	Introduction	1
Chapter 2	Description of Samples and Population	2
Chapter 3	Probability and the Binomial Distribution	6
Chapter 4	The Normal Distribution	10
Chapter 5	Sampling Distributions	18
Unit I		25
Chapter 6	Confidence Intervals	27
Chapter 7	Comparison of Two Independent Samples	31
Chapter 8	Comparison of Paired Samples	40
Unit II		45
Chapter 9	Categorical Data: One Sample Distributions	46
Chapter 10	Categorical Data: Relationships	53
Unit III		62
Chapter 11	Comparing the Means of Many Independent Samples	63
Chapter 12	Linear Regression and Correlation	70
Unit IV		78
Chapter 13	A Summary of Inference Methods	79

CHAPTER 1
Introduction

1.2.3 The acupuncturist expects acupuncture to work better than aspirin, so she or he is apt to "see" more improvement in someone given acupuncture than in someone given aspirin -- even if the two groups are truly equivalent to each other in their response to treatment.

1.3.1 (a) Cluster sampling. The three clinics are the three clusters.

(b) Simple random sampling.

(c) Stratified random sampling. The strata are the altitudes.

(d) Simple random sampling.

(e) Stratified random sampling. The three breed sizes are the strata.

1.3.2 (a) The sample is nonrandom and likely nonrepresentative of the general population because it consists of (1) volunteers from (2) nightclubs. (i) The social anxiety level of people who attend nightclubs is likely lower than the social anxiety level of the general public. (ii) A better sampling strategy would be to recruit subjects from across the population.

(b) Bias arises from the data being collected only on rainy days. (i) Water pollution readings in the stream might be lower when rain water is mixed in with regular stream water. (ii) A better method would be to sample during all types of weather.

(c) Recording observations only when random coordinates are within a tree canopy will induce bias. (i) Trees with large canopies are more likely to be included in the sample than are other trees, but canopy size is probably related to tree radius, resulting in a sample average that is too large. (ii) A better sampling method would be to measure whichever tree trunk is closest to randomly choose coordinates, but even this produces bias in favor of large trees. In order to avoid the bias, number all of the trees within a region and drawn random numbers to select trees to measure. Geographical regions could be used as strata within a stratified sampling plan of this type.

(d) Fish caught by a single vessel on one day are not a random sample. (i) If the vessel is in a region that has not been fished recently and thus contains large fish, for example, then the sample average will be too large. (ii) To avoid this bias, use randomly chosen fishing vessels on randomly chosen days.

CHAPTER 2
Description of Samples and Populations

2.1.2 (a) i) Height and weight
ii) Continuous variables
iii) A child
iv) 37

(b) i) Blood type and cholesterol level
ii) Blood type is categorical, cholesterol level is continuous
iii) A person
iv) 129

2.2.1 (a) There is no single correct answer. One possibility is:

Molar width	Frequency (no. specimens)
[5.4, 5.6)	1
[5.6, 5.8)	5
[5.8, 6.0)	7
[6.0, 6.2)	12
[6.2, 6.4)	8
[6.4, 6.6)	2
[6.6, 6.8)	1
Total	36

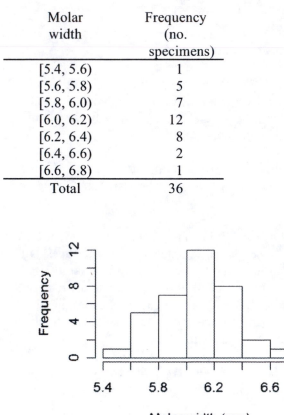

(b) The distribution is fairly symmetric.

2.2.7 There is no single correct answer. One possibility is

Glucose (%)	Frequency (no. of dogs)
70-74	3
75-79	5
80-84	10

85-89	5
90-94	2
95-99	2
100-104	1
105-109	1
110-114	0
115-119	1
120-124	0
125-129	0
130-134	1
Total	31

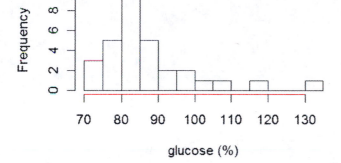

2.3.1 Any sample with $\Sigma y_i = 100$ would be a correct answer. For example: 18, 19, 20, 21, 22.

2.3.5 $\bar{y} = 293.8$ mg/dl; median = 283 mg/dl.

2.3.6 $\bar{y} = 309$ mg/dl; median = 292 mg/dl.

2.3.11 The median is the average of the 18th and 19th largest values. There are 18 values less than or equal to 10 and 18 values that are greater than or equal to 11. Thus, the median is

$$\frac{10+11}{2} = 10.5 \text{ piglets.}$$

2.3.13 The distribution is fairly symmetric so the mean and median are roughly equal. It appears that half of the distribution is below 50 and half is above 50. Thus, mean ≈ median ≈ 50.

2.4.2 (a) The median is the average of the 9th and 10th largest observations. The ordered list of the data is

4.1 5.2 6.8 7.3 7.4 7.8 7.8 8.4 8.7 9.7 9.9 10.6 10.7 11.9 12.7 14.2 14.5 18.8

Thus, the median is $\dfrac{8.7+9.7}{2} = 9.2$.

To find Q_1 we consider only the lower half of the data set:

4.1 5.2 6.8 7.3 7.4 7.8 7.8 8.4 8.7 9.7

Q_1 is the median of this half (i.e., the 5th largest value), which is 7.4.

To find Q_3 we consider only the upper half of the data set:

9.7 9.9 10.6 10.7 11.9 12.7 14.2 14.5 18.8.

Q_3 is the median of this half (i.e., the 5th largest value in this list), which is 11.9.

(b) IQR = Q_3 - Q_1 = 11.9 - 7.4 = 4.5.

(c) Upper fence = Q_3 + 1.5 × IQR = 11.9 + 6.75 = 18.65.

(d)

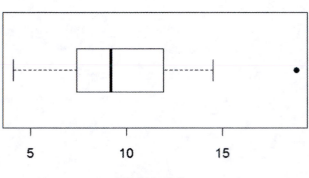

MAO Activity

2.6.1 (a) $\bar{y} = 15$, $\sum(y_i - \bar{y})^2 = 18$, $s = \sqrt{18/3} = 2.45$.

(b) $\bar{y} = 35$, $\sum(y_i - \bar{y})^2 = 44$, $s = \sqrt{44/4} = 3.32$.

(c) $\bar{y} = 1$, $\sum(y_i - \bar{y})^2 = 24$, $s = \sqrt{24/3} = 2.83$.

(d) $\bar{y} = 3$, $\sum(y_i - \bar{y})^2 = 28$, $s = \sqrt{28/4} = 2.65$.

2.6.4 $\bar{y} = 33.10$ lb; $s = 3.444$ lb.

2.6.9 (a) $\bar{y} \pm s$ is 32.23 ± 8.07, or 24.16 to 40.30; this interval contains 10/15 or 67% of the observations.

(b) $\bar{y} \pm 2s$ is 16.09 to 48.37; this interval contains 15/15 or 100% of the observations.

2.6.14 Coefficient of variation = $\dfrac{s}{\bar{y}} = \dfrac{6.8}{166.3} = 0.04$ or 4%.

2.6.15 The mean is about 45. The length of the interval that covers the middle 95% of the data is approximately equal to 70 - 20 = 50. An estimate of s is (length of interval)/4 = 50/4 ≈ 12.

2.7.1 $y' = (y - 7)*100$. Thus, the mean of y' is $(\bar{y} - 7)*100 = (7.373 - 7)*100 = 37.3$. The SD of y' is $s \times 100 = 0.129*100 = 12.9$.

2.S.8 The bars in the histogram that correspond to observations less than 45 represent roughly one-third of the total area. Thus, about 30% of the observations are less than 45.

2.S.15 (a) n = 119, so the median is the 60th largest observation. There are 32 observations less than or equal to 37 and 44 observations less than or equal to 38. Thus, the median is 38.

(b) The first quartile is the 30th largest observation, which is 36. The third quartile is the 90th largest observation, which is 41.

(c)

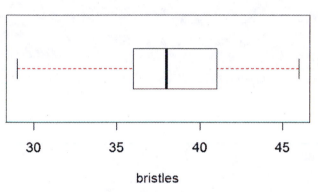

bristles

(d) The mean is 38.45 and the SD is 3.20. Thus, the interval $\bar{y} \pm s$ is 38.45 ± 3.20, which is 35.25 to 41.65. This interval includes 36, 27, 28, 29, 40, and 41. The number of flies with 36 to 41 bristles is $11 + 12 + 18 + 13 + 10 + 15 = 79$. Thus, the percentage of observations that fall within one standard deviation of the mean is $79/119*100\% = 0.664*100\% = 66.4\%$.

(e) The quartiles are 36 and 41. Between 36 and 41 (inclusive) there are 79 observations out of 119 total so 66.4%.

CHAPTER 3
Probability and the Binomial Distribution

3.2.1 (a) In the population, 51% of the fish have 21 vertebrae. Thus, $\Pr\{Y = 21\} = 0.51$.

   (b) In the population, the percentage of fish with 22 or fewer vertebrae is $3 + 51 + 40 = 94\%$. Thus, $\Pr\{Y \leq 22\} = 0.94$.

   (c) In the population, the percentage of fish with more than 21 vertebrae is $40 + 6 = 46\%$. Thus, $\Pr\{Y > 21\} = 0.46$.

   (d) In the population, the percentage of fish with no more than 21 vertebrae is $3 + 51 = 54\%$. Thus, $\Pr\{Y \leq 21\} = 0.54$.

3.2.6 (a)

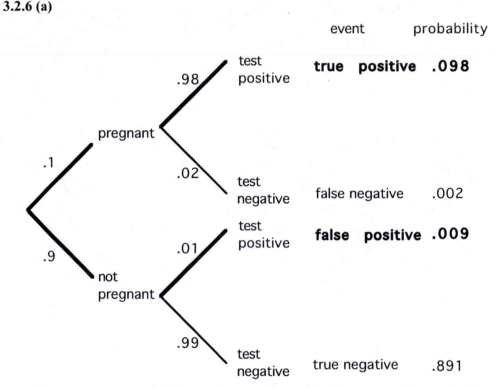

    There are two ways to test positive. A true positive happens with probability $(0.1)(0.98) = 0.098$. A false positive happens with probability $(0.9)(0.01) = 0.009$. Thus, $\Pr\{\text{test positive}\} = 0.098 + 0.009 = 0.107$.

   (b) Using the same reasoning as in part (a),
   $\Pr\{\text{test positive}\} = (0.05)(0.98) + (0.95)(0.01) = 0.049 + 0.095 = 0.0585$.

3.2.7 (a) $0.098/0.107 = 0.916$

   (b) $0.049/0.0585 = 0.838$

3.3.1 (a) $1213/6549 = 0.1852 \approx 0.185$

(b) $247/2115 = 0.1168 \approx 0.117$

(c) No; the probability of a person being a smoker depends on whether or not the person has high income, since the answers to (a) and (b) differ.

3.4.3 (a) $\Pr\{20 < Y < 30\} = 0.41 + 0.21 = 0.62$

(b) $0.41 + 0.21 + 0.03 = 0.65$

(c) $0.01 + 0.34 = 0.35$

3.5.5 $(0)(0.343) + (1)(0.441) + (2)(0.189) + (3)(0.027) = 0.9$

3.5.6 $\text{VAR}(Y) = (0 - 0.9)^2(0.343) + (1 - 0.9)^2(0.441) + (2 - 0.9)^2(0.189) + (3 - 0.9)^2(0.027) = 0.63$.

Thus, the standard deviation is $\sqrt{0.63} = 0.794$.

3.6.8 On average, there are 105 males to every 100 females. Thus, $\Pr\{\text{male}\} = \dfrac{105}{205}$ and $\Pr\{\text{female}\} = \dfrac{100}{205}$. To use the binomial distribution, we arbitrarily identify "success" as "female."

(a) We have $n = 4$ and $p = \dfrac{100}{205}$. To find the probability of 2 males and 2 females, we set $j = 2$, so $n - j = 2$. The binomial formula gives $\Pr\{2 \text{ males and } 2 \text{ females}\} =$

$$_4C_2 \left(\frac{100}{205}\right)^2 \left(\frac{105}{205}\right)^2 = 0.3746.$$

(b) To find the probability of 4 males, we set $j = 0$, so $n - j = 4$. The binomial formula gives

$$\Pr\{4 \text{ males}\} = {_4C_0} \left(\frac{100}{205}\right)^0 \left(\frac{105}{205}\right)^4 = (1)(1) \left(\frac{105}{205}\right)^4 = 0.0688.$$

(c) The condition that all four infants are the same sex can be satisfied two ways: All four could be male or all four could be female. The probability that all four are male has been computed in part (b) to be 0.0688. To find the probability that all four are females, we set $j = 4$, so $n - j = 0$.

$$\Pr\{4 \text{ females}\} = {_4C_4} \left(\frac{100}{205}\right)^4 \left(\frac{105}{205}\right)^0 = (1)(1) \left(\frac{100}{205}\right)^4 = 0.0566.$$

Thus, we find that $\Pr\{\text{all four are the same sex}\} = 0.0566 + 0.0688 = 0.1254$.

3.6.11 (a) $0.75^6 = 0.1780$

(b) $1 - 0.1780 = 0.8220$

3.7.1 The first step is to determine the best-fitting value for $p = \Pr\{\text{boy}\}$. The total number of children in all the families is

$$(6)(72,069) = 432,414.$$

The number of boys is

$$(0)(1,096) + (1)(6,233) + \ldots + (6)(1,579) = 222,638.$$

Thus, the value of p that fits the data best is

$$p = \frac{222638}{432414} = 0.514872321.$$

To compute the probabilities of various sex ratios, we apply the binomial formula with $n = 6$ and $p = 0.514872321$. Then we multiply each probability by 72,069 to obtain the expected frequency:

Number of boys (j)	Expected frequency		
0	$(72,069)(1)(1 - p)^6$	=	939.5
1	$(72,069)(6)(p)(1 - p)^5$	=	5,982.5
2	$(72,069)(15)(p^2)(1 - p)^4$	=	15,873.1
3	$(72,069)(20)(p^3)(1 - p)^3$	=	22,461.8
4	$(72,069)(15)(p^4)(1 - p)^2$	=	17,879.3
5	$(72,069)(6)(p^5)(1 - p)^1$	=	7,590.2
6	$(72,069)(1)(p^6)$	=	1,342.6

The following table compares the observed and expected frequencies:

Number of boys	Number of girls	Observed frequency	Expected frequency	Sign of (Obs - Exp)
0	6	1,096	939.5	+
1	5	6,233	5,982.5	+
2	4	15,700	15,873.1	-
3	3	22,221	22,461.8	-
4	2	17,332	17,879.3	-
5	1	7,908	7,590.2	+
6	0	1,579	1,342.6	+
		72,069	72,069.0	

We note that there is reasonable agreement between the observed and expected frequencies. However, the observed frequencies exceed the expected frequencies for the preponderantly unisex siblingships (those with 0, 1, 5, or 6 boys), whereas the observed frequencies are less than the expected frequencies for the more balanced siblingships (2, 3, or 4 boys). This pattern is similar to that seen in Example 3.7.1.

3.S.3 The probability that a square has no centipedes is the relative frequency of squares with zero centipedes. Thus, $\Pr\{\text{no centipedes}\} = 0.45$.

To apply the binomial formula, we identify "success" as "no centipedes." Then n = 5 and p = 0.45. To find the probability that three squares have centipedes and two do not, we set j = 2, so n - j = 3.

Pr{3 with, 2 without} = $_5C_2(0.45^2)(0.55^3)$ = $10(0.45^2)(0.55^3)$ = 0.3369.

3.S.7 (a) $1 - 0.99^{100}$ = 0.6340

(b) $1 - 0.99^n \geq 0.95$, so $n \geq \log(0.05)/\log(0.99)$, so $n \geq 299$.

3.S.10 (a) To use the binomial distribution here, let us identify "success" as "blood pressure > 140." Then, p = Pr{blood pressure > 140} = 0.25 + 0.09 + 0.04 = 0.38, which is the area under the density curve beyond 140 mm Hg.

The number of trials is n = 4. To find the probability that all four men have blood pressure higher than 140 mm Hg, we set j = 4, so n - j = 0.

Pr{all four have blood pressure > 140} = $_4C_4p^4(1-p)^0$ = $(1)(0.38^4)(1)$ = 0.0209.

(b) Pr{three have blood pressure > 140} = $_4C_3p^3(1-p)^1$ = $(4)(0.38^3)(0.62^1)$ = 0.1361.

3.S.12 (a) The men is (50)(0.09) = 4.5

(b) The SD is $\sqrt{(50)(0.09)(0.91)}$ = 2.02

CHAPTER 4
The Normal Distribution

4.3.3 $\mu = 1400$; $\sigma = 100$.

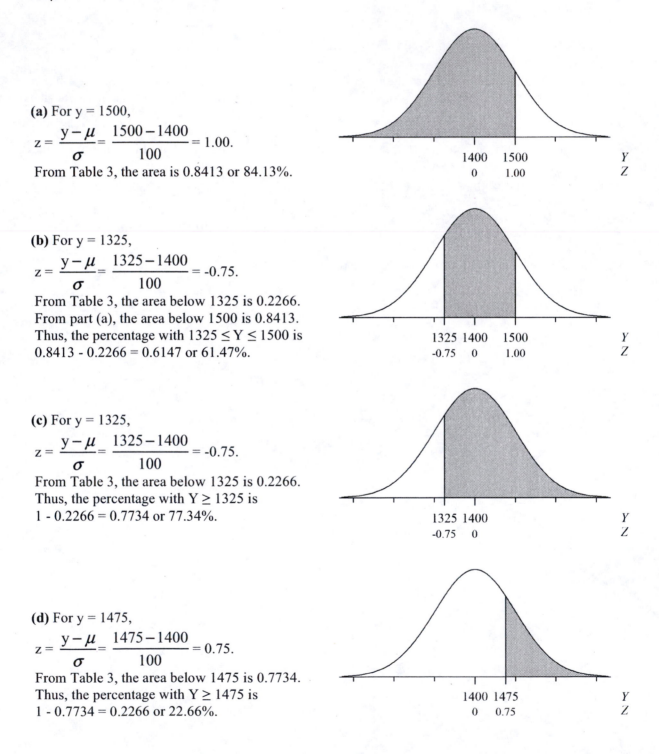

(a) For y = 1500,

$$z = \frac{y - \mu}{\sigma} = \frac{1500 - 1400}{100} = 1.00.$$

From Table 3, the area is 0.8413 or 84.13%.

(b) For y = 1325,

$$z = \frac{y - \mu}{\sigma} = \frac{1325 - 1400}{100} = -0.75.$$

From Table 3, the area below 1325 is 0.2266.
From part (a), the area below 1500 is 0.8413.
Thus, the percentage with $1325 \leq Y \leq 1500$ is
0.8413 - 0.2266 = 0.6147 or 61.47%.

(c) For y = 1325,

$$z = \frac{y - \mu}{\sigma} = \frac{1325 - 1400}{100} = -0.75.$$

From Table 3, the area below 1325 is 0.2266.
Thus, the percentage with $Y \geq 1325$ is
1 - 0.2266 = 0.7734 or 77.34%.

(d) For y = 1475,

$$z = \frac{y - \mu}{\sigma} = \frac{1475 - 1400}{100} = 0.75.$$

From Table 3, the area below 1475 is 0.7734.
Thus, the percentage with $Y \geq 1475$ is
1 - 0.7734 = 0.2266 or 22.66%.

(e) For y = 1600,

$$z = \frac{y - \mu}{\sigma} = \frac{1600 - 1400}{100} = 2.00.$$

From Table 3, the area below 1600 is 0.9772.
In part (d) we found that the area below 1475 is 0.7734.
Thus, the percentage with $1475 \leq Y \leq 1600$ is 0.9772 - 0.7734 = 0.2038 or 20.38%.

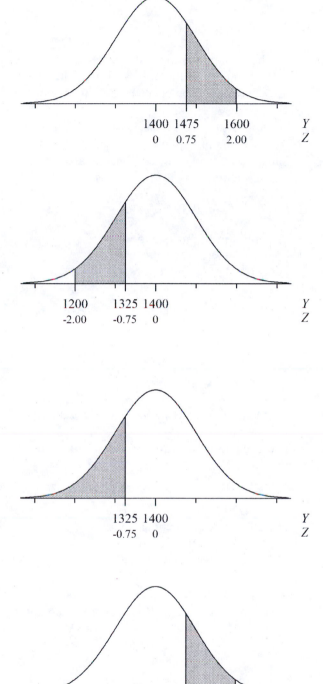

1400	1475	1600	Y
0	0.75	2.00	Z

(f) For y = 1200,

$$z = \frac{y - \mu}{\sigma} = \frac{1200 - 1400}{100} = -2.00.$$

From Table 3, the area below 1200 is 0.0228.
In part (c) we found that the area below 1325 is 0.2266.
Thus, the percentage with $1200 \leq Y \leq 1325$ is 0.2266 - 0.0028 = 0.2038 or 20.38%.

1200	1325	1400	Y
-2.00	-0.75	0	Z

4.3.4 $\mu = 1400$; $\sigma = 100$.

(a) For y = 1325,

$$z = \frac{y - \mu}{\sigma} = \frac{1325 - 1400}{100} = -0.75.$$

From Table 3, the area is 0.2266 or 22.66%.

1325	1400	Y
-0.75	0	Z

(b) This is the same as part (e) of Exercise 4.3.3.
For y = 1600,

$$z = \frac{y - \mu}{\sigma} = \frac{1600 - 1400}{100} = 2.00.$$

From Table 3, the area below 1600 is 0.9772.
For y = 1475,

$$z = \frac{y - \mu}{\sigma} = \frac{1475 - 1400}{100} = 0.75.$$

From Table 3, the area below 1475 is 0.7734.
Thus, the percentage with $1475 \leq Y \leq 1600$ is 0.9772 - 0.7734 = 0.2038 or 20.38%.

1400	1475	1600	Y
0	0.75	2.00	Z

4.3.8 $\mu = 88$; $\sigma = 7$.

(a) The 65th percentile is the value that is larger than 65% of the observations. Thus, the area under the curve below this value is 0.65.

In Table 3, the area closest to 0.65 is 0.6517, which corresponds to $z = 0.39$.

Thus, the 65th percentile y^* must satisfy the equation

$$z = \frac{y^* - \mu}{\sigma} \text{ or } 0.39 = \frac{y^* - 88}{7}.$$

The solution of this equation is

$$y^* = (7)(0.39) + 88 = 90.7.$$

Thus, the 65th percentile is 90.7 lb.

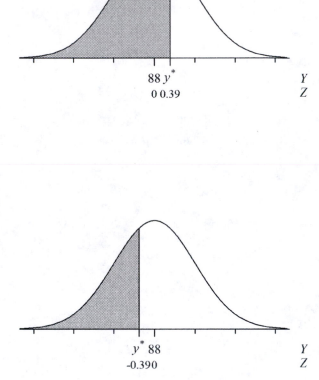

(b) The 35th percentile is the value that is larger than 35% of the observations. Thus, the area under the curve below this value is 0.35.

In Table 3, the area closest to 0.35 is 0.3483, which corresponds to $z = -0.39$.

(Note that this is the negative of the value found in part (a).)

Thus, the 35th percentile y^* must satisfy the equation

$$z = \frac{y^* - \mu}{\sigma} \text{ or } -0.39 = \frac{y^* - 88}{7}.$$

The solution of this equation is

$$y^* = (7)(-0.39) + 88 = 85.3.$$

Thus, the 35th percentile is 85.3 lb.

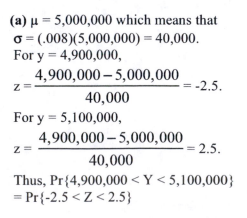

4.3.12 The distribution of readings is a normal distribution with mean μ (the true value) and standard deviation $\sigma = .008\mu$ (0.8% of the true value).

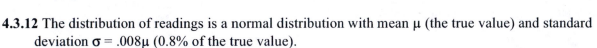

(a) $\mu = 5,000,000$ which means that

$\sigma = (.008)(5,000,000) = 40,000$.

For $y = 4,900,000$,

$$z = \frac{4,900,000 - 5,000,000}{40,000} = -2.5.$$

For $y = 5,100,000$,

$$z = \frac{4,900,000 - 5,000,000}{40,000} = 2.5.$$

Thus, $\Pr\{4,900,000 < Y < 5,100,000\}$

$= \Pr\{-2.5 < Z < 2.5\}$

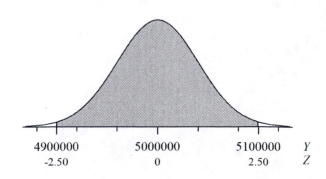

= 0.9938 - 0.0062 = 0.9876 or 98.76%.

(b) $\Pr\{0.98\mu < Y < 1.02\mu\} = \Pr\{\dfrac{0.98\mu - \mu}{0.008\mu} < \dfrac{Y - \mu}{\sigma} < \dfrac{1.02\mu - \mu}{0.008\mu}\}$

= Pr{-2.5 < Z < 2.5} = 0.9938 - 0.0062 = 0.9876 or 98.76%

(c) A specimen reading Y differs from the correct value by 2% or more if it does <u>not</u> satisfy 0.98μ < Y < 1.02μ. Using the answer from part (b), this probability is 1 - 0.9876 = 0.0124 or 1.24%.

4.4.3 The histograms below are histograms of the data used to generate the normal probability plots. Your sketched histogram may look different than these, but should contain similar features: (a) heavy tails, and (b) left skew.

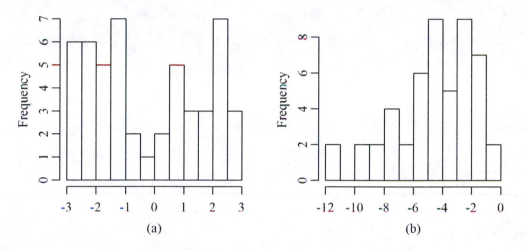

(a)   (b)

4.4.8 (a) No, it does not seem reasonable to believe that these data came from a normal population, since the P-value = 0.039 is small (indeed, it is less than 0.05).   Thus, we have moderate evidence that the population from which this data comes is abnormal.

(b) Yes, it seems reasonable to believe that these data came from a normal population, since the P-value = 0.770 ≥ 0.10. That is, there is not even weak evidence for abnormality in the population.

4.S.4 μ = 145; σ = 22.

(a) For y = 100,

$z = \dfrac{y - \mu}{\sigma} = \dfrac{100 - 145}{22} = -2.05.$

From Table 3, the area below -2.05 is 0.0202.
Thus, Pr{Y ≥ 100} = 1 - 0.0202 = 0.9798 or 97.98%.

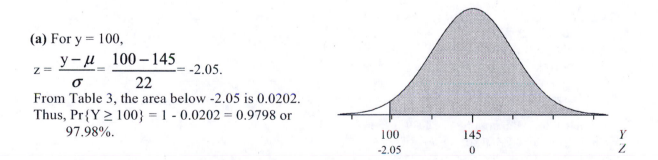

(b) For y = 120,

$$z = \frac{y - \mu}{\sigma} = \frac{120 - 145}{22} = -1.14.$$

From Table 3, the area below -1.14 is 0.1271.
Thus, $\Pr\{Y \le 120\} = 0.1271$ or 12.71%.

(c) For y = 120,

$$z = \frac{y - \mu}{\sigma} = \frac{120 - 145}{22} = -1.14.$$

From Table 3, the area below -1.14 is 0.1271.
For y = 150,

$$z = \frac{y - \mu}{\sigma} = \frac{150 - 145}{22} = 0.23.$$

From Table 3, the area below 0.23 is 0.5910.
Thus, $\Pr\{120 \le Y \le 150\} =$
0.5910 - 0.1271 = 0.4639 or 46.39%.

(d) For y = 100,

$$z = \frac{y - \mu}{\sigma} = \frac{100 - 145}{22} = -2.05.$$

From Table 3, the area below -2.05 is 0.0202.
For y = 120,

$$z = \frac{y - \mu}{\sigma} = \frac{120 - 145}{22} = -1.14.$$

From Table 3, the area below -1.14 is 0.1271.
Thus, $\Pr\{100 \le Y \le 120\} =$
0.1271 - 0.0202 = 0.1069 or 10.69%.

(e) For y = 150,

$$z = \frac{y - \mu}{\sigma} = \frac{150 - 145}{22} = 0.23.$$

From Table 3, the area below 0.23 is 0.5910.
For y = 180,

$$z = \frac{y - \mu}{\sigma} = \frac{180 - 145}{22} = 1.59.$$

From Table 3, the area below 1.59 is 0.9441.
Thus, $\Pr\{150 \le Y \le 180\} =$
0.9441 - 0.5910 = 0.3531 or 35.31%.

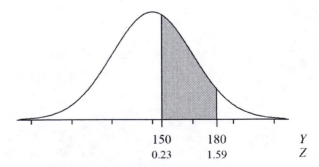

(f) For y = 180,

$$z = \frac{y-\mu}{\sigma} = \frac{180-145}{22} = 1.59.$$

From Table 3, the area below 1.59 is 0.9441.
Thus, $\Pr\{Y \geq 150\} = 1 - 0.9441 = 0.0559$ or
  5.59%.

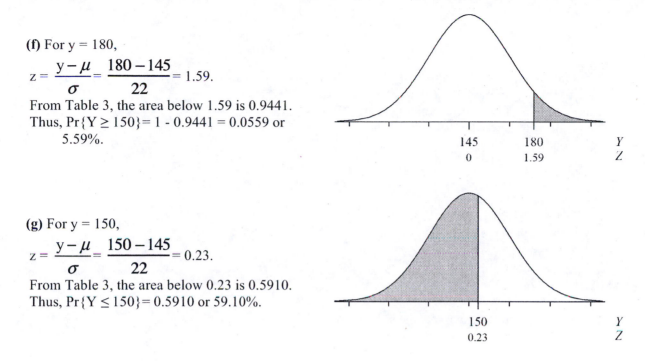

(g) For y = 150,

$$z = \frac{y-\mu}{\sigma} = \frac{150-145}{22} = 0.23.$$

From Table 3, the area below 0.23 is 0.5910.
Thus, $\Pr\{Y \leq 150\} = 0.5910$ or 59.10%.

4.S.5 $\mu = 145$; $\sigma = 22$.

If none of the plants is more than 150 cm tall, then all of the plants are less than or equal to 150 cm tall.

For y = 150,

$$z = \frac{y-\mu}{\sigma} = \frac{150-145}{22} = 0.23.$$

From Table 3, the area below 0.23 is 0.5910.
Thus, $\Pr\{Y \leq 150\} = 0.5910$ or 59.10%.

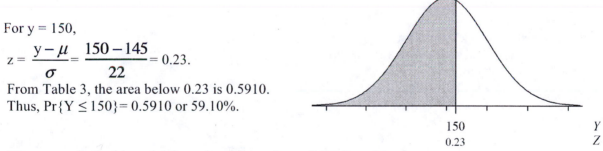

We can apply the binomial formula with n = 4, p = 0.5910, and j = 4.
Thus, Pr{none more than 150 cm tall}
= Pr{all less than or equal to 150 cm tall}
= $_4C_4 0.5910^4 = 0.122$.

4.S.6 $\mu = 145$; $\sigma = 22$.

The 90th percentile is the value that is larger than 90% of the distribution. In Table 3, the area closest to .9 is .8997, corresponding to z = 1.28. Thus, the 90th percentile y* satisfies the equation

$$1.28 = \frac{y^* - 145}{22}.$$

The solution of this equation is
    y* = (22)(1.28) + 145 = 173.2.
Thus, the 90th percentile of the distribution is 173.2 cm.

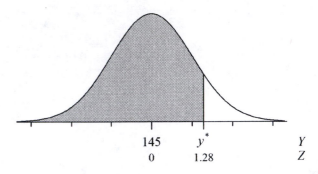

4.S.8 $\mu = 145$; $\sigma = 22$.

(a) We wish to find z* such that the shaded area is 0.95. This means that the area in the left tail is 0.025. In Table 3, the area 0.025 corresponds to z = -1.96, which is -z*. Likewise, the area 0.975 corresponds to z = 1.96. Thus, z* = 1.96.

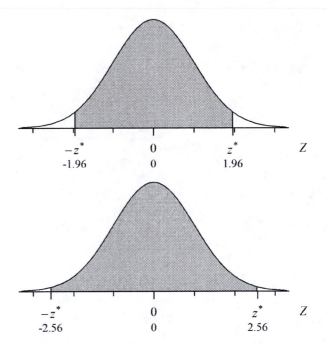

(b) We wish to find z* such that the shaded area is 0.99. This means that the area in the left tail is 0.005. In Table 3, the area 0.0005 corresponds to z = -2.58, which is -z*. Likewise, the area 0.995 corresponds to z = 2.56. Thus, z* = 2.58.

4.S.15 The distribution of readings is a normal distribution with mean μ (the true concentration) and standard deviation σ. A reading of 40 or more is considered "unusually high." Suppose that μ = 35 and σ = 4.

For y = 40,

$$z = \frac{40-35}{4} = 1.25.$$

From Table 3, the area below 1.25 is 0.8944, which means that the area above 1.25 is 1 - 0.8944 = 0.1056. Thus,

Pr{specimen is flagged as "unusually high} = 0.1056.

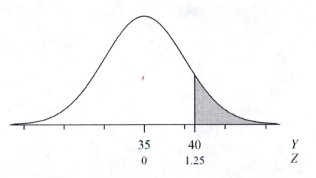

4.S.18 Pr{0 < Y < 15} = 0.7549 - 0.2546 = 0.5003. Thus we expect (400)(0.5003), or about 200 observations to fall between 0 and 15.

CHAPTER 5
Sampling Distributions

5.1.2 There are three possible total costs in between $80,000 and $125,000 when sampling n = 3 patients as listed in Table 5.1.2: $95,000, $105,000, and $120,000. Thus,

$$\Pr\{\text{total cost} > \$125,000\} = \Pr\{\text{total cost} = \$95,000\} + \Pr\{\text{total cost} = \$105,000\}$$
$$+ \Pr\{\text{total cost} = \$120,000\}$$
$$= 12/64 + 8/64 + 3/64$$
$$= 23/64 = 0.3594$$

5.2.4

(a) In the population, $\mu = 155$ and $\sigma = 27$.
For y = 165,

$$z = \frac{y - \mu}{\sigma} = \frac{165 - 155}{27} = 0.37..$$

From Table 3, the area below 0.37 is 06443.
For y = 145,

$$z = \frac{y - \mu}{\sigma} = \frac{145 - 155}{27} = -0.37.$$

From Table 3, the area below -0.37 is 0.3557.
Thus, the percentage with $145 \le y \le 165$
is $0.6443 - 0.3557 = 0.2886$, or 28.86%.

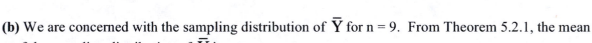

(b) We are concerned with the sampling distribution of $\overline{Y}$ for n = 9. From Theorem 5.2.1, the mean of the sampling distribution of $\overline{Y}$ is

$$\mu_{\overline{Y}} = \mu = 155,$$

the standard deviation is

$$\sigma_{\overline{Y}} = \frac{\sigma}{\sqrt{n}} = \frac{27}{\sqrt{9}} = 9,$$

and the shape of the distribution is normal because the population distribution is normal (part 3a of Theorem 5.2.1).

We need to find the shaded area in the figure.
For $\overline{y} = 165$,

$$z = \frac{\overline{y} - \mu_{\overline{Y}}}{\sigma_{\overline{Y}}} = \frac{165 - 155}{9} = 1.11.$$

From Table 3, the area below 1.11 is 0.8665.
For $\overline{y} = 145$,

$$z = \frac{\overline{y} - \mu_{\overline{Y}}}{\sigma_{\overline{Y}}} = \frac{145 - 155}{9} = -1.11.$$

From Table 3, the area below -1.11 is 0.1335.
Thus, the percentage with $145 \le \overline{y} \le 165$
is $0.8665 - 0.135 = 0.7330$, or 73.30%.

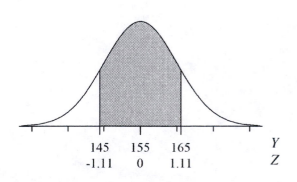

(c) The probability of an event can be interpreted as the long-run relative frequency of occurrence of the event (Section 3.2). Thus, the question in part (c) is just a rephrasing of the question in part (b). It follows from part (b) that

$\Pr\{145 \le \overline{Y} \le 165\} = 0.7330.$

5.2.6 (a) $\mu = 3000$; $\sigma = 400$.

The event E occurs if $\overline{Y}$ is between 2900 and 3100. We are concerned with the sampling distribution of $\overline{Y}$ for n = 15. From Theorem 5.2.1, the mean of the sampling distribution of $\overline{Y}$ is

$\mu_{\overline{Y}} = \mu = 3000,$

the standard deviation is

$$\sigma_{\overline{Y}} = \frac{\sigma}{\sqrt{n}} = \frac{400}{\sqrt{15}} = 103.3,$$

and the shape of the distribution is normal because the population distribution is normal (part 3a of Theorem 5.2.1).
For $\overline{y} = 3100$,

$z = \dfrac{\overline{y} - \mu_{\overline{Y}}}{\sigma_{\overline{Y}}} = \dfrac{3100 - 3000}{103.3} = 0.97.$

From Table 3, the area below 0.97 is 0.8340.
For $\overline{y} = 2900$,

$z = \dfrac{\overline{y} - \mu_{\overline{Y}}}{\sigma_{\overline{Y}}} = \dfrac{2900 - 3000}{103.3} = -0.97.$

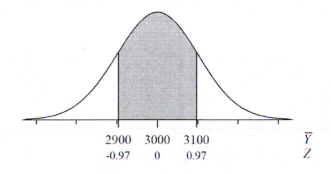

From Table 3, the area below -0.97 is 0.1660.
Thus, $\Pr\{2900 \le \overline{Y} \le 3100\}$
$= \Pr\{E\} = .8340 - .1660 = .6680.$

(b) n = 60; $\sigma_{\overline{Y}} = 400 / \sqrt{60} = 51.64$

$z = \dfrac{\pm 100}{51.64} = \pm 1.94$; Table 3 gives 0.9738 and 0.0262, so $\Pr\{E\} = 0.9738 - 0.0262 = 0.9476$.

(c) As n increases, $\Pr\{E\}$ increases.

5.2.10 (a) In the population, 65.68% of the fish are between 51 and 60 mm long. To find the probability that four randomly chosen fish are all between 51 and 60 mm long, we let "success" be "between 51 and 60 mm long" and use the binomial distribution with n = 4 and p = 0.6568, as follows:

$\Pr\{\text{all 4 are between 51 and 60}\} = {}_4C_4p^4(1 - p)^0 = (1)0.6568^4(1) = 0.1861.$

(b) The mean length of four randomly chosen fish is $\overline{Y}$. Thus, we are concerned with the sampling distribution of $\overline{Y}$ for a sample of size n = 4 from a population with $\mu = 54$ and $\sigma = 4.5$. From Theorem 5.2.1, the mean of the sampling distribution of $\overline{Y}$ is

$\mu_{\overline{Y}} = \mu = 54,$

the standard deviation is

$$\sigma_{\overline{Y}} = \frac{\sigma}{\sqrt{n}} = \frac{4.5}{\sqrt{4}} = 2.25,$$

and the shape of the distribution is normal because the population distribution is normal (part 3a of Theorem 5.2.1).

For $\overline{y} = 60$,

$$z = \frac{\overline{y} - \mu_{\overline{Y}}}{\sigma_{\overline{Y}}} = \frac{60 - 54}{2.25} = 2.67.$$

From Table 3, the area below 2.67 is 0.9962.
For $\overline{y} = 51$,

$$z = \frac{\overline{y} - \mu_{\overline{Y}}}{\sigma_{\overline{Y}}} = \frac{51 - 54}{2.25} = -1.33.$$

From Table 3, the area below -1.33 is 0.0918.
Thus, $\Pr\{51 \le \overline{Y} \le 60\}$
$= 0.9962 - 0.0918 = 0.9044.$

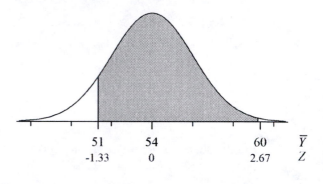

5.2.13 The distribution of repeated assays of the patient's specimen is a normal distribution with mean $\mu = 35$ (the true concentration) and standard deviation $\sigma = 4$.

(a) The result of a single assay is like a random observation Y from the population of assays. A value $Y \ge 40$ will be flagged as "unusually high." For $y = 40$,

$$z = \frac{y - \mu}{\sigma} = \frac{40 - 35}{4} = 1.25.$$

From Table 3, the area below 1.25 is 0.8944, so the area beyond 1.25 is
$1 - 0.8944 = 0.1056.$

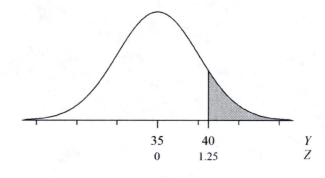

Thus, $\Pr\{\text{specimen will be flagged as "unusually high"}\} = 0.1056.$

(b) The reported value is the mean of three independent assays, which is like the mean $\overline{Y}$ of a sample of size n = 3 from the population of assays. A value $\overline{Y} \ge 40$ will be flagged as "unusually high." We are concerned with the sampling distribution of $\overline{Y}$ for a sample of size n = 3 from a population with mean $\mu = 35$ and standard deviation $\sigma = 4$. From Theorem 5.2.1, the mean of the sampling distribution of $\overline{Y}$ is

$$\mu_{\overline{Y}} = \mu = 35,$$

the standard deviation is

$$\sigma_{\overline{Y}} = \frac{\sigma}{\sqrt{n}} = \frac{4}{\sqrt{3}} = 2.309,$$

and the shape of the distribution is normal because the population distribution is normal (part 3a of Theorem 5.2.1).

For $\bar{y} = 40$,

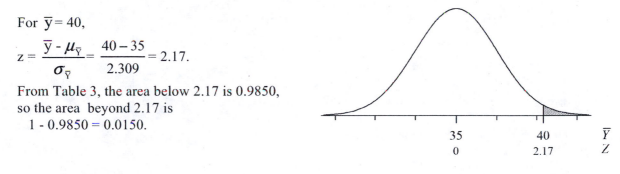

$$z = \frac{\bar{y} - \mu_{\bar{Y}}}{\sigma_{\bar{Y}}} = \frac{40 - 35}{2.309} = 2.17.$$

From Table 3, the area below 2.17 is 0.9850,
so the area beyond 2.17 is
$1 - 0.9850 = 0.0150.$

Thus, Pr{mean of three assays will be flagged as "unusually high"}$= 1 - 0.9850 = 0.0150.$

5.2.15 (a) $\mu_{\bar{Y}} = \mu = 41.5.$

(b) $\sigma_{\bar{Y}} = 4.7 / \sqrt{4} = 2.35$

5.3.1 For each thrust, the probability is 0.9 that the thrust is good and the probability is 0.1 that the thrust is fumbled. Letting "success" = "good thrust," and assuming that the thrusts are independent, we apply the binomial formula with $n = 4$ and $p = 0.9$.

(a) The area under the first peak is approximately equal to the probability that all four thrusts are good. To find this probability, we set $j = 4$; thus, the area is approximately
$$_4C_4 p^4 (1 - p)^0 = (1)(0.9^4)(1) = 0.66.$$

(b) The area under the second peak is approximately equal to the probability that three thrusts are good and one is fumbled. To find this probability, we set $j = 3$; thus, the area is approximately
$$_4C_3 p^3 (1 - p)^1 = (4)(0.9^3)(0.1) = 0.29.$$

5.4.2 Letting "success" = "heads," the probability of ten heads and ten tails is determined by the binomial distribution with $n = 20$ and $p = 0.5$.

(a) We apply the binomial formula with $j = 10$:
$$\Pr\{10 \text{ heads}, 10 \text{ tails}\} = _{20}C_{10} p^{10} (1 - p)^{10} = (184,756)(0.5^{10})(0.5^{10}) = 0.1762.$$

(b) According to part (a) of Theorem 5.4.1, the binomial distribution can be approximated by a normal distribution with
mean $= np = (20)(0.5) = 10$
and
standard deviation $= \sqrt{np(1-p)} = \sqrt{(20)(0.5)(0.5)} = 2.236.$
Applying continuity correction, we wish to find the area under the normal curve between $10 - 0.5 = 9.5$ and $10 + 0.5 = 10.5$.

The desired area is shaded in the figure.

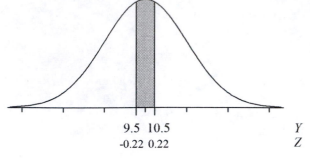

The boundary 10.5 corresponds to

$$z = \frac{10.5 - 10}{2.236} = 0.22.$$

From Table 3, the area below 0.22 is 0.5871.

The boundary 9.5 corresponds to

$$z = \frac{9.5 - 10}{2.236} = -0.22.$$

From Table 3, the area below -0.22 is 0.4129.

Thus, the normal approximation to the binomial probability is
Pr{10 heads, 10 tails} ≈ 0.5871 - 0.4129 = 0.1742.

5.4.5 (a) Because p = 0.12, the event that $\hat{p}$ will be within ±0.03 of p is the event

$$0.09 \le \hat{p} \le 0.15,$$

which, if n = 100, is equivalent to the event

$$9 \le \text{number of success} \le 15.$$

Letting "success" = "oral contraceptive user," the probability of this event is determined by the binomial distribution with
    mean = np = (100)(0.12) = 12
and

standard deviation = $\sqrt{np(1-p)} = \sqrt{100(0.12)(0.88)} = 3.250.$

Applying continuity correction, we wish to find the area under the normal curve between
    9 - 0.5 = 8.5 and 15 + 0.5 = 15.5.

The desired area is shaded in the figure.

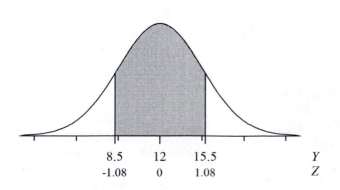

The boundary 15.5 corresponds to

$$z = \frac{15.5 - 12}{3.250} = 1.08.$$

From Table 3, the area below 1.08 is 0.8599.

The boundary 8.5 corresponds to

$$z = \frac{8.5 - 12}{3.250} = -1.08.$$

From Table 3, the area below -1.08 is 0.1401.

Thus, the normal approximation to the binomial probability is
Pr{$\hat{p}$ will be within ±0.03 of p} ≈ 0.8599 - 0.1401 = 0.7198.

(Note: An alternative method of solution is to use part (b) of Theorem 5.4.1 rather than part (a).)

(b) With n = 200, $\hat{p}$ is within ±0.03 of p if and only if the number of successes is between (200)(0.09) = 18 and (200)(0.15) = 30. The mean is (200)(0.12) = 24 and the standard deviation is $\sqrt{200(0.12)(0.88)} = 4.60$.

Applying continuity correction, we wish to find the area under the normal curve between 18 - 0.5 = 17.5 and 30 + 0.5 = 30.5.

$$z = \frac{30.5 - 24}{4.60} = 1.41; \text{ Table 3 gives } 0.9207.$$

$$z = \frac{17.5 - 24}{4.60} = -1.41; \text{ Table 3 gives } 0.0793.$$

0.9207 - 0.0783 = 0.8414.

5.4.7 p ±0.05 is 0.25 to 0.35.

The normal approximation to the sampling distribution of $\hat{p}$ has mean p = 0.3 and standard deviation

$$\sqrt{\frac{p(1-p)}{n}} = \sqrt{\frac{(0.3)(0.7)}{400}} = 0.02291.$$

$$z = \frac{0.35 - 0.3}{0.02291} = 2.18; \text{ Table 3 gives } 0.9854.$$

$$z = \frac{0.25 - 0.3}{0.02291} = -2.18; \text{ Table 3 gives } 0.0146.$$

0.9854 - 0.0146 = 0.9708.

5.S.1 μ = 88; σ = 7.

We are concerned with the sampling distribution of $\overline{Y}$ for n = 5. From Theorem 5.2.1, the mean of the sampling distribution of $\overline{Y}$ is

$$\mu_{\overline{Y}} = \mu = 88,$$

the standard deviation is

$$\sigma_{\overline{Y}} = \frac{\sigma}{\sqrt{n}} = \frac{7}{\sqrt{5}} = 3.13,$$

and the shape of the distribution is normal because the population distribution is normal (part 3a of Theorem 5.2.1).

For $\bar{y}$ = 90,

$$z = \frac{\bar{y} - \mu_{\bar{Y}}}{\sigma_{\bar{Y}}} = \frac{90 - 88}{3.13} = 0.64.$$

From Table 3, the area below 0.64 is 0.7389.
Thus, $\Pr\{\bar{Y} > 90\} = 1 - 0.7389 = 0.2611$.

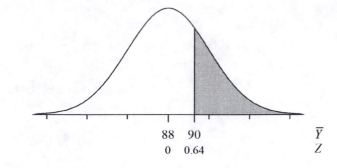

5.S.9 $\mu = 8.3$; $\sigma = 1.7$.

If the total weight of 10 mice is 90 gm, then their mean weight is

$$\frac{90}{10} = 9.0 \text{ gm.}$$

Thus, we wish to find the percentage of litters for which $\bar{y} \geq 9.0$ gm. We are concerned with the

sampling distribution of $\bar{Y}$ for n = 10. From Theorem 5.2.1, the mean of the sampling distribution

of $\bar{Y}$ is

$$\mu_{\bar{Y}} = \mu = 8.3,$$

the standard deviation is

$$\sigma_{\bar{Y}} = \frac{\sigma}{\sqrt{n}} = \frac{1.7}{\sqrt{10}} = 0.538,$$

and the shape of the distribution is normal because the population distribution is normal (part 3a of
Theorem 5.2.1).

We need to find the shaded area in the figure.
For $\bar{y} = 9.0$,

$$z = \frac{\bar{y} - \mu_{\bar{Y}}}{\sigma_{\bar{Y}}} = \frac{9.0 - 8.3}{0.538} = 1.30.$$

From Table 3, the area below 1.30 is 0.9032.
Thus, the percentage with $\bar{y} \geq 9.0$ is

$$1 - 0.9032 = 0.0968, \text{ or } 9.68\%.$$

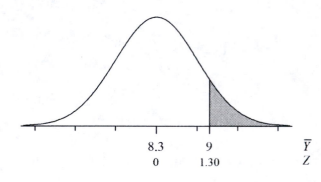

UNIT I

I.1 (a) For Graph II, the mean – 2(SD) is 1.7 – 2.2, which is negative. But precipitation totals can't be negative; moreover we know that the minimum is 0.3. Thus, Graph II can't be right. Graph III is even worse. Thus, the answer must be Graph I (which fits having min ≈ mean – 1*SD).

(b) The median is *less than* the mean because the distribution is skewed to the right. The long right tail pulls up the mean.

I.4 The sampling distribution of the sample percentage is the distribution of $\hat{P}$, the sample percentage of mice weighing more then 26 gm, as it varies from one sample of 20 mice to another in repeated samples.

I.5 (a) $\Pr\{16 < Y < 23\} = \Pr\left\{\dfrac{16-20}{5} < Z < \dfrac{23-20}{5}\right\} = \Pr\{-0.80 < Z < 0.60\} = 0.514.$

(b) $\bar{Y}$ has a normal distribution with mean 20 and SD $5/\sqrt{5} = 2.24$. $\Pr\{16 < \bar{Y} < 23\} =$

$\Pr\left\{\dfrac{16-20}{5/\sqrt{5}} < Z < \dfrac{23-20}{2.24}\right\} = \Pr\{-1.79 < Z < 1.34\} = 0.9099 - 0.0367 = 0.873.$

I.10 $\Pr\{Y \geq 54\} = \Pr\left\{Z \geq \dfrac{54 - 100*9/16}{\sqrt{100*(9/16)*(7/16)}}\right\} = \Pr\left\{Z \geq \dfrac{54 - 56.25}{4.96}\right\} = \Pr(Z \geq -0.45) = 0.674.$

I.15 (a) $z_{.15} = 1.04$

$y = 53.3 + 1.04 \times 14.6 = 68.48$ mg/L

(b) 85th percentile

(c) Note that this question pertains to the sampling distribution of the sample mean for samples of size 7. Thus, we wish to compute $\Pr(\bar{Y} > 70.4)$ where $\bar{Y} \sim N(\mu_{\bar{Y}} = 53.3, \sigma_{\bar{Y}} = \frac{14.6}{\sqrt{7}})$.

$z = \dfrac{70.4 - 53.3}{14.6/\sqrt{7}} \approx 3.10$

$\Pr(Z > 3.10) = 1 - \Pr(Z < 3.10) = 1 - 0.9990 = 0.0010$

This probability is very small; hence, it would be very unusual to observe a sample of seven rats with a sample mean thiocyanate level above 53.3 mg/L.

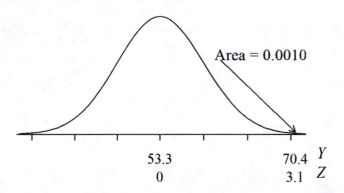

CHAPTER 6
Confidence Intervals

6.2.1 (a) $\bar{y} = 1269$; $s = 145$; $n = 8$.

The standard error of the mean is

$$SE_{\bar{Y}} = \frac{s}{\sqrt{n}} = \frac{145}{\sqrt{8}} = 51.3 \text{ ng/gm}.$$

(b) $\bar{y} = 1269$; $s = 145$; $n = 30$.

The standard error of the mean is

$$SE_{\bar{Y}} = \frac{s}{\sqrt{n}} = \frac{145}{\sqrt{30}} = 26.5 \text{ ng/gm}.$$

6.3.3 (a) $\bar{y} = 31.720$ mg; $s = 8.729$ mg; $n = 5$.

The standard error of the mean is

$$SE_{\bar{Y}} = \frac{s}{\sqrt{n}} = \frac{8.7}{\sqrt{5}} = 3.89 \approx 3.9 \text{ mg}.$$

(b) The degrees of freedom are n - 1 = 5 - 1 = 4. The critical value is $t_{0.05} = 2.132$. The 90% confidence interval for μ is

$$\bar{y} \pm t_{0.05} \frac{s}{\sqrt{n}}$$

$$31.7 \pm 2.132 \left(\frac{8.7}{\sqrt{5}} \right)$$

$(23.4, 40.0)$ or $23.4 < \mu < 40.0$ mg.

6.3.11 (a) $\bar{y} = 13.0$; $s = 12.4$; $n = 10$.

The degrees of freedom are n - 1 = 10 - 1 = 9. The critical value is $t_{0.025} = 2.262$. The 95% confidence interval for μ is

$$\bar{y} \pm t_{0.025} \frac{s}{\sqrt{n}}$$

$$13.0 \pm 2.262 \left(\frac{12.4}{\sqrt{10}} \right)$$

$(4.1, 21.9)$ or $4.1 < \mu < 21.9$ pg/ml.

(b) We are 95% confident that the average drop in HBE levels from January to May in the population of all participants in physical fitness programs like the one in the study is between 4.1 and 21.9 pg/ml.

(c) From the computed 95% confidence interval we have evidence that the mean drop in HBE is at least 4.1 pg/ml. So yes, there is compelling evidence that the HBE level tends to be lower in May than January, on average.

6.3.15 $\bar{y} = 1.20$; $s = .14$; $n = 50$.

The degrees of freedom are 50 - 1 = 49. From Table 4 with df = 50 (the df value closest to 49) we find that $t_{0.05} = 1.676$. The 90% confidence interval for μ is

$$\bar{y} \pm t_{0.05} \frac{s}{\sqrt{n}}$$

$$1.20 \pm 1.676 \left(\frac{0.14}{\sqrt{50}} \right)$$

(1.17, 1.23) or $1.17 < \mu < 1.23$ mm.

6.3.20 1 - 0.0025 = 0.9975. In Table 3, an area of 0.9975 corresponds to z = 2.81. A t distribution with df = ∞ is a normal distribution; thus, $t_{0.0025} = 2.81$ when df = ∞.

6.4.2 We use the inequality

$$\frac{\text{Guessed SD}}{\sqrt{n}} \leq \text{Desired SE.}$$

In this case, the desired SE is 3 mg/dl and the guessed SD is 40 mg/dl. Thus, the inequality is

$$\frac{40}{\sqrt{n}} \leq 3 \quad \text{or} \quad \frac{40}{3} \leq \sqrt{n} \text{ which means that } n \geq 177.8, \text{ so a sample of } n = 178 \text{ men is needed.}$$

6.4.6 Guessed SD = 1.5 in

The SE should be no more than .25 in, so n must satisfy

$$\frac{1.5}{\sqrt{n}} \leq .25$$

which yields $n \geq 36$.

6.5.1 The fact that the mean is less than the SD casts doubt on the condition that the population is normal, for the following reason. In a normal population, about 15% of the observations fall more than one SD below the mean, whereas this sample cannot have any observations that far below the mean because $\bar{y}$ - s is negative and the observed variable (serum SGOT) cannot be negative.

6.5.6 In this experiment the treatments have been assigned to the flasks, so the entire flasks are the independent units. Though randomly drawn from within each flask, the three aliquots are not independent from one another in the larger context of this study: the aliquots from the same flask will be more similar to each other than aliquots from different flasks. Due to this hierarchical sampling, the SE as described would be an underestimate of the true SE. The SE as computed violates Condition 1(b). (One correct analysis would be to compute the mean density of the three aliquots from each flask, and then use these values as the independent measurements for analysis)

6.6.1 We first find the standard error of each mean.

$$SE_1 = SE_{\bar{Y}_1} = \frac{s_1}{\sqrt{n_1}} = \frac{4.3}{\sqrt{6}} = 1.755.$$

$$SE_2 = SE_{\bar{Y}_2} = \frac{s_2}{\sqrt{n_2}} = \frac{5.7}{\sqrt{12}} = 1.645.$$

$$SE_{(\bar{Y}_1 - \bar{Y}_2)} = \sqrt{SE_1^2 + SE_2^2} = \sqrt{1.755^2 + 1.645^2} = 2.41.$$

6.6.7 $SE_1 = \dfrac{2.4}{\sqrt{49}} = 0.343;\ SE_2 = \dfrac{2.0}{\sqrt{52}} = 0.277.$

$$SE_{(\bar{Y}_1 - \bar{Y}_2)} = \sqrt{SE_1^2 + SE_2^2} = \sqrt{0.343^2 + 0.277^2} = 0.44.$$

6.7.1 Let 1 denote conventional treatment and let 2 denote Coblation.

$$\bar{y}_1 = 4.3;\ SE_1 = 2.8 / \sqrt{49} = 0.4.$$

$$\bar{y}_2 = 1.9;\ SE_2 = 1.8 / \sqrt{52} = 0.25.$$

The standard error of the difference is $SE_{(\bar{y}_1 + \bar{y}_2)} = \sqrt{0.4^2 + 0.25^2} = 0.47.$

The critical value $t_{0.025}$ is determined from Student's t distribution with df = 81.1. Using df = 80 (the nearest value given in Table 4), we find that $t_{80,0.025} = 1.990$.

The 95% confidence interval is

$$\bar{y}_1 - \bar{y}_2 \pm t_{0.025} SE_{(\bar{Y}_1 - \bar{Y}_2)}$$

$$(4.3 - 1.9) \pm (1.990)(0.47).$$

6.7.6 (a) Let 1 denote antibiotic and let 2 denote control.

$$SE_{(\bar{Y}_1 - \bar{Y}_2)} = \sqrt{\frac{10^2}{10} + \frac{8^2}{10}} = 4.050.$$

$$(25 - 23) \pm (1.740)(4.050) \qquad\qquad (\text{using df} = 17)$$

$$(-5.0, 9.0) \text{ or } -5 < \mu_1 - \mu_2 < 9 \text{ sec.}$$

(b) For the interval to be valid, the sampling distribution of $\bar{Y}_1 - \bar{Y}_2$ must be approximately normal. This will occur when either the data itself comes from a normally distributed population, or when the sample sizes are large (via the Central Limit Theorem). Since the sample sizes are not particularly large (10 and 10) in this case, we must trust that the data itself follows a normal distribution. (Note that such faith might not be wise. One should have a good reason for this belief, or else poor decisions based on the data are possible.)

(c) We are 90% confident that the population mean prothrombin time for rats treated with an antibiotic (μ_1) is smaller than that for control rats (μ_2) by an amount that might be as much as 5 seconds or is larger than that for control rats (μ_2) by an amount that might be as large as 9 seconds.

6.S.2 (a) $\bar{y} = 2.275$; s = 0.238. The standard error of the mean is

$$SE_{\bar{y}} = \frac{s}{\sqrt{n}} = \frac{0.238}{\sqrt{8}} = 0.084 \text{ mm.}$$

(b) From Table 4 with df = n - 1 = 7, we find that $t_{0.025} = 2.365$.
The 95% confidence interval for μ is

$$\bar{y} \pm t_{0.025} \frac{s}{\sqrt{n}}$$

$$2.275 \pm (2.365)(.084)$$

$$(2.08, 2.47) \text{ or } 2.08 < \mu < 2.47 \text{ mm.}$$

(c) μ is the population mean stem diameter of plants of Tetrastichon wheat three weeks after flowering.

6.S.4 We use the inequality

$$\frac{\text{Guessed SD}}{\sqrt{n}} \leq \text{Desired SE.}$$

In this case the desired SE is 0.03 mm and the guessed SD (from Exercise 6.46) is 0.238 mm. Thus, the inequality is

$$\frac{0.238}{\sqrt{n}} \leq 0.03 \text{ or } \frac{0.238}{0.03} \leq \sqrt{n}, \text{ so } 7.933^2 \leq n, \text{ which means that } n \geq 62.9.$$

Thus, the experiment should include 63 plants.

6.S.9 (a) We must be able to view the data as a random sample of observations from a large population, the observations in the sample must be independent of each other, and the population distribution must be approximately normal. (Note, however, that because the sample size (n = 28) is not very small, some non-normality of the population distribution would be acceptable.)

(b) The shape of the histogram is an estimate of the shape of the population distribution. Thus, the histogram can be used to check the normality condition of the population

(c) If twin births were included, the independence of the observations would be questionable, because birthweights of the members of a twin pair are likely to be dependent.

CHAPTER 7
Comparison of Two Independent Samples

7.1.2 (b) Nine of the ten randomizations (in bold above) yield differences that are at least as large in magnitude as -2.73.

7.2.1 (a) The observed t statistic is

$$t_s = \frac{\bar{y}_1 - \bar{y}_2}{SE_{(\bar{Y}_1 - \bar{Y}_2)}} = \frac{735 - 854}{38} = -3.13.$$

From Table 4 with df = 4, we find the critical values $t_{0.02} = 2.999$ and $t_{0.01} = 3.747$. Because t_s is between $t_{0.01}$ and $t_{0.02}$, the P-value must be between .02 and .04. Thus, the P-value is bracketed as $0.02 < \text{P-value} < 0.04$.

(b) The observed t statistic is

$$t_s = \frac{\bar{y}_1 - \bar{y}_2}{SE_{(\bar{Y}_1 - \bar{Y}_2)}} = \frac{5.3 - 5.0}{0.24} = 1.25.$$

From Table 4 with df = 12, we find the critical values $t_{0.20} = .873$ and $t_{0.10} = 1.356$. Because t_s is between $t_{0.10}$ and $t_{0.20}$, the P-value must be between 0.20 and 0.40. Thus, the P-value is bracketed as $0.20 < \text{P-value} < 0.40$.

(c) The observed t statistic is

$$t_s = \frac{\bar{y}_1 - \bar{y}_2}{SE_{(\bar{Y}_1 - \bar{Y}_2)}} = \frac{36 - 30}{1.3} = 4.62.$$

From Table 4 with df = 30, we find the critical value $t_{0.0005} = 3.646$. Because t_s is greater than $t_{0.0005}$, the P-value must be less than 0.001. Thus, the P-value is bracketed as $\text{P-value} < 0.001$.

7.2.3 (a) $0.085 < 0.10$, which means that the P-value is less than α. Thus, we reject H_0.

(b) $0.065 > 0.05$, which means that the P-value is greater than α. Thus, we do not reject H_0.

(c) Table 4 gives $t_{19,0.005} = 2.861$ and $t_{19,0.0005} = 3.883$, so $0.001 < \text{P-value} < 0.01$. Since $P < \alpha$, we reject H_0.

(d) Table 4 gives $t_{12,0.05} = 1.782$ and $t_{12,0.04} = 1.912$, so $.008 < \text{P-value} < 0.10$. Since $P > \alpha$, we do not reject H_0.

Remark concerning tests of hypotheses The answer to a hypothesis testing exercise includes verbal statements of the hypotheses and a verbal statement of the conclusion from the test. In phrasing these statements, we have tried to capture the essence of the biological question being addressed; nevertheless the statements are necessarily oversimplified and they gloss over many issues that in reality might be quite important. For instance, the hypotheses and conclusion may refer to a causal connection between treatment and response; in reality the validity of such a causal interpretation usually depends on a number of factors related to the design of the investigation (such as unbiased allocation of animals to treatment groups) and to the specific experimental procedures (such as the accuracy of assays or measurement techniques). In short, the student should be aware that the verbal statements are intended to clarify the *statistical* concepts; their *biological* content may be open to question.

7.2.7 (a) The null and alternative hypotheses are

$$H_0: \mu_1 = \mu_2$$
$$H_A: \mu_1 \neq \mu_2$$

where 1 denotes heart disease and 2 denotes control. These hypotheses may be stated as

H_0: Mean serotonin concentration is the same in heart patients and in controls
H_A: Mean serotonin concentration is not the same in heart patients and in controls

The test statistic is

$$t_s = \frac{\bar{y}_1 - \bar{y}_2}{SE_{(\bar{Y}_1 - \bar{Y}_2)}} = \frac{3840 - 5310}{1064} = -1.38.$$

From Table 4 with df = 14, we find the critical values $t_{0.10} = 1.345$ and $t_{0.05} = 1.761$. Thus, the P-value is bracketed as 0.10 < P-value < 0.20.
Since the P-value is greater than α (0.05), H_0 is not rejected. There is insufficient evidence (0.10 < P < 0.20) to conclude that serotonin levels are different in heart patients than in controls.

7.2.11 The null and alternative hypotheses are

$$H_0: \mu_1 = \mu_2$$
$$H_A: \mu_1 \neq \mu_2$$

where 1 denotes flooded and 2 denotes control. These hypotheses may be stated as

H_0: Flooding has no effect on ATP
H_A: Flooding has some effect on ATP

The standard error of the difference is

$$SE_{(\bar{Y}_1 - \bar{Y}_2)} = \sqrt{\frac{0.184^2}{4} + \frac{0.241^2}{4}} = 0.1516.$$

The test statistic is

$$t_s = \frac{\bar{y}_1 - \bar{y}_2}{SE_{(\bar{Y}_1 - \bar{Y}_2)}} = \frac{1.190 - 1.785}{0.1516} = -3.92.$$

From Table 4 with df = $n_1 + n_2 - 2 = 6$ (Formula (6.7.1) yields df = 5.6), we find the critical values $t_{0.005} = 3.707$ and $t_{0.0005} = 5.959$. Thus, the P-value is bracketed as 0.001 < P-value < 0.01.
Since the P-value is less than α (0.05), we reject H_0. There is sufficient evidence (0.001 < P < 0.01) to conclude that flooding tends to lower ATP in birch seedlings.

7.3.4 If we reject H_0 (i.e., if the drug is approved) then we eliminate the possibility of a Type II error. (But by rejecting H_0 we may have made a Type I error.)

7.3.6 Yes; because zero is outside of the confidence interval, we know that the P-value is less than 0.05, so we reject the hypothesis that $\mu_1 - \mu_2 = 0$.

7.4.1 No, this does not mean that living in Arizona exacerbates breathing problems. To determine this, we would need to know whether breathing problems get better or worse for people in Arizona. In fact, people with respiratory problems often move to Arizona because the dry air is good for them. This would explain the association between living in Arizona and having breathing problems.

7.4.4 (a) The explanatory variable is coffee consumption rate.

(b) The response variable is coronary heart disease (present or absent).

(c) The observational units are subjects (i.e., the 1,040 persons).

7.5.1 (a) The observed t statistic is

$$t_s = \frac{\overline{y}_1 - \overline{y}_2}{SE_{(\overline{Y}_1 - \overline{Y}_2)}} = \frac{10.8 - 10.5}{0.23} = 1.30.$$

To check the directionality of the data, we note that $\overline{y}_1 > \overline{y}_2$. Thus, the data do deviate from H_0 in the direction ($\mu_1 > \mu_2$) specified by H_A, and therefore the one-tailed P-value is the area under the t curve beyond 1.30.

From Table 4 with df = 18, we find the critical values $t_{0.20} = 0.862$ and $t_{0.10} = 1.330$. Because t_s is between $t_{0.10}$ and $t_{0.20}$, the one-tailed P-value must be between 0.10 and 0.20. Thus, the P-value is bracketed as $0.10 < P < 0.20$.

(b) The observed t statistic is

$$t_s = \frac{\overline{y}_1 - \overline{y}_2}{SE_{(\overline{Y}_1 - \overline{Y}_2)}} = \frac{750 - 730}{11} = 1.82.$$

To check the directionality of the data, we note that $\overline{y}_1 > \overline{y}_2$. Thus, the data do deviate from H_0 in the direction ($\mu_1 > \mu_2$) specified by H_A, and therefore the one-tailed P-value is the area under the t curve beyond 1.82.

From Table 4 with df = 140 (the closest value to 180), we find the critical values $t_{0.04} = 1.763$ and $t_{0.03} = 1.896$. Because t_s is between $t_{0.03}$ and $t_{0.04}$, the one-tailed P-value must be between 0.03 and 0.04. Thus, the P-value is bracketed as $0.03 < P < 0.04$.

7.5.3 (a) Yes. t_s is positive, as predicted by H_A. Thus, the P-value is the area under the t curve beyond 3.75. With df = 19, Table 4 gives $t_{0.005} = 2.861$ and $t_{0.0005} = 3.883$. Thus, $0.0005 < P\text{-value} < 0.005$. Since $P < \alpha$, we reject H_0.

(b) Yes. t_s is positive, as predicted by H_A. Thus, the P-value is the area under the t curve beyond 2.6.

With df = 5, Table 4 gives $t_{0.025} = 2.571$ and $t_{0.02} = 2.757$. Thus, $0.02 < $ P-value < 0.025. Since $P < \alpha$, we reject H_0.

(c) Yes. t_s is positive, as predicted by H_A. Thus, the P-value is the area under the t curve beyond 2.1. With df = 7, Table 4 gives $t_{0.04} = 2.046$ and $t_{0.03} = 2.241$. Thus, $0.03 < $ P-value < 0.04. Since $P < \alpha$, we reject H_0.

(d) No. t_s is positive, as predicted by H_A. Thus, the P-value is the area under the t curve beyond 1.8. With df = 7, Table 4 gives $t_{0.10} = 1.415$ and $t_{0.05} = 1.895$. Thus, $0.05 < $ P-value < 0.10. Since $P > \alpha$, we do not reject H_0.

7.5.9 The null and alternative hypotheses are

$$H_0: \mu_1 = \mu_2$$
$$H_A: \mu_1 < \mu_2$$

where 1 denotes wounded and 2 denotes control. These hypotheses may be stated as

H_0: Wounding the plant has no effect on larval growth
H_A: Wounding the plant tends to diminish larval growth

To check the directionality of the data, we note that $\bar{y}_1 < \bar{y}_2$. Thus, the data do deviate from H_0 in the direction $(\mu_1 < \mu_2)$ specified by H_A. We proceed to calculate the test statistic.
The standard error of the difference is

$$SE_{(\bar{Y}_1 - \bar{Y}_2)} = \sqrt{\frac{9.02^2}{16} + \frac{11.14^2}{18}} = 3.46.$$

The test statistic is

$$t_s = \frac{\bar{y}_1 - \bar{y}_2}{SE_{(\bar{Y}_1 - \bar{Y}_2)}} = \frac{28.66 - 37.96}{3.46} = -2.69.$$

From Table 4 with df = 16 + 18 -2 = 32 $\approx$ 30 (Formula (6.7.1) yields df = 31.8), we find the critical values $t_{0.01} = 2.457$ and $t_{0.005} = 2.750$. Thus, the P-value is bracketed as $0.005 < $ P-value < 0.01.
Since the P-value is less than α (0.05), we reject H_0. There is sufficient evidence $(0.005 < P < 0.01)$ to conclude that wounding the plant tends to diminish larval growth.

7.5.10 (a) The null and alternative hypotheses are

$$H_0: \mu_1 = \mu_2$$
$$H_A: \mu_1 > \mu_2$$

where 1 denotes drug and 2 denotes placebo. These hypotheses may be stated as

H_0: The drug is not effective

H_A: The drug is effective

To check the directionality of the data, we note that $\bar{y}_1 > \bar{y}_2$. Thus, the data do deviate from H_0 in the direction ($\mu_1 > \mu_2$) specified by H_A. We proceed to calculate the test statistic.

The standard error of the difference is

$$SE_{(\bar{Y}_1 - \bar{Y}_2)} = \sqrt{\frac{12.05^2}{25} + \frac{13.78^2}{25}} = 3.66.$$

The test statistic is

$$t_s = \frac{\bar{y}_1 - \bar{y}_2}{SE_{(\bar{Y}_1 - \bar{Y}_2)}} = \frac{31.96 - 25.32}{3.66} = 1.81.$$

From Table 4 with df = 25 + 25 - 2 = 48 ≈ 50 (Formula (6.7.1) yields df = 47.2), we find the critical values $t_{0.04} = 1.787$ and $t_{0.03} = 1.924$. Thus, the P-value is bracketed as

0.03 < P-value < 0.04.

Since the P-value is less than α (0.05), we reject H_0. There is sufficient evidence (0.03 < P < 0.04) to conclude that the drug is effective at increasing pain relief.

(b) The only change in the calculations from part (a) would be that the one-tailed area would be doubled if the alternative were nondirectional. Thus, the p-value would be between 0.06 and 0.08 and at $\alpha = 0.05$ we would *not* reject H_0.

7.6.4 The mean difference in concentration of coumaric acid is $\mu_1 - \mu_2$, where 1 denotes dark and 2 denotes photoperiod. We construct a 95% confidence interval for $\mu_1 - \mu_2$.

$$SE_{(\bar{Y}_1 - \bar{Y}_2)} = \sqrt{\frac{21^2}{4} + \frac{27^2}{4}} = 17.103.$$

The critical value $t_{0.025}$ is found from Student's t distribution with df = $n_1 + n_2 - 2 = 6$. (Formula (6.7.1) gives df = 5.7.) From Table 4, we find $t_{6,0.025} = 2.447$. The 95% confidence interval is

$$\bar{y}_1 - \bar{y}_2 \pm t_{0.025} SE_{(\bar{Y}_1 + \bar{Y}_2)}$$

$$(106 - 102) \pm (2.447)(17.103)$$

$$(-37.9, 45.9) \text{ or } -37.9 < \mu_1 - \mu_2 < 45.9 \text{ nmol/gm.}$$

The difference could be larger than 20 nmol/gm or much smaller, so the data do not indicate whether the difference is "important."

7.6.6 Using the larger SD the sample effect size is computed as

$$\frac{55.3 - 53.3}{6.1} = 0.3279.$$

7.7.1 From the preliminary data, we obtain 0.3 cm as a guess of σ.
  (a) If the true difference is 0.25 cm, then the effect size is

$$\frac{|\mu_1 - \mu_2|}{\sigma} = \frac{0.25}{0.3} = 0.83.$$

We consult Table 5 for a two-tailed test at $\alpha = 0.05$ and an effect size of $0.83 \approx 0.85$; to achieve power 0.80, Table 5 recommends $n = 23$.

(b) If the true difference is 0.5 cm, then the effect size is

$$\frac{|\mu_1 - \mu_2|}{\sigma} = \frac{0.5}{0.3} = 1.67.$$

We consult Table 5 for a two-tailed test at $\alpha = 0.05$ and an effect size of $1.67 \approx 1.7$; to achieve power 0.95, Table 5 recommends $n = 11$.

7.7.4 We need to find n to achieve a power of 0.9. The effect size is

$$\frac{|\mu_1 - \mu_2|}{\sigma} = \frac{81 - 75}{11} = 0.55.$$

We consult Table 5.
  (a) For a two-tailed test at $\alpha = 0.05$, Table 5 gives $n = 71$.
  (b) For a two-tailed test at $\alpha = 0.01$, Table 5 gives $n = 101$.
  (c) For a one-tailed test at $\alpha = 0.05$, Table 5 gives $n = 58$.

7.7.6 The effect size is

$$\frac{|\mu_1 - \mu_2|}{\sigma} = \frac{4}{10} = 0.4.$$

From Table 5 we find that, for a one-tailed test with $n = 35$ at significance level $\alpha = 0.05$, the power is 0.50 if the effect size is 0.4. Thus, the probability that Jones will reject H_0 is equal to 0.50.

7.10.1 We consult Table 6 with $n = 7$ and $n' = 5$.

(a) $U_s = 26$. From Table 6, the smallest critical value is 27, which corresponds to a nondirectional P-value of 0.149. Since $U_s < 27$, it follows that $P > 0.149$.

(b) $U_s = 30$. From Table 6, the P-value is 0.048.

(c) $U_s = 35$. From Table 6, the P-value is 0.0025.

7.10.3 (a) The null and alternative hypotheses are

H_0: Toluene has no effect on dopamine in rat striatum

H_A: Toluene has some effect on dopamine in rat striatum

Let 1 denote toluene and let 2 denote control. The ordered arrays of observations are as follows:

Y_1:   1911   2314 2464   2781 2803 3420

Y_2: 1397 1803 1820 1843   1990   2539

For the K_1 count, we note that there are four Y_2's less than the first Y_1; there are five Y_2's less than the second Y_1; there are five Y_2's less than the third Y_1; and there are six Y_2's less than the fourth, fifth, and sixth Y_1. Thus,

$$K_1 = 4 + 5 + 5 + 6 + 6 + 6 = 32.$$

For the K_2 count, we note that there are no Y_1's less than the first, second, third, or fourth Y_2; there is one Y_1 less than the fifth Y_2; and there are three Y_1's less than the sixth Y_2. Thus,

$$K_2 = 0 + 0 + 0 + 0 + 1 + 3 = 4.$$

To check the counts, we verify that

$$K_1 + K_2 = 32 + 4 = 36 = (6)(6) = (n_1)(n_2).$$

The Wilcoxon-Mann-Whitney test statistic is the larger of the two counts K_1 and K_2; thus $U_s = 32$.

Looking in Table 6 under n = 6 and n' = 6, we find that for a nondirectional alternative, the entry in the 0.05 column is 31, for which the P-value is 0.041, and the entry in the 0.025 column is 33, for which the P-value is 0.015. Thus, the P-value is bracketed as

$$0.015 < \text{P-value} < 0.041.$$

At significance level $\alpha = 0.05$, we reject H_0, since $P < 0.041 < 0.05$. We note that K_1 is larger than K_2, which indicates a tendency for the Y_1's to be larger than the Y_2's. Thus, there is sufficient evidence ($0.015 < P < 0.041$) to conclude that toluene increases dopamine in rat striatum.

7.S.2 The null and alternative hypotheses are

H_0: Mean platelet calcium is the same in people with high blood pressure as in people with normal blood pressure ($\mu_1 = \mu_2$)

H_A: Mean platelet calcium is different in people with high blood pressure than in people with normal blood pressure ($\mu_1 \neq \mu_2$)

The standard error of the difference is

$$SE_{(\bar{Y}_1 - \bar{Y}_2)} = \sqrt{\frac{16.1^2}{38} + \frac{31.7^2}{45}} = 5.399.$$

The test statistic is

$$t_s = \frac{\bar{y}_1 - \bar{y}_2}{SE_{(\bar{Y}_1 - \bar{Y}_2)}} = \frac{168.2 - 107.9}{5.399} = 11.2.$$

From Table 4 with df $= 45 + 38 - 2 = 81 \approx 80$, we find the critical value $t_{0.0005} = 3.416$. The tail area is doubled for the nondirectional test. Thus, the P-value is bracketed as $P < 0.001$. (Formula (6.7.1) yields df $= 67.5$, but the P-value is still bracketed as $P < 0.001$.)

Since the P-value is less than α (0.01), we reject H_0. There is sufficient evidence ($P < 0.001$) to conclude that mean platelet calcium is higher in people with high blood pressure than in people with normal blood pressure.

7.S.4 No; the t test is valid because the sample sizes are rather large.

7.S.8 The null and alternative hypotheses are

H_0: Stress has no effect on growth
H_A: Stress tends to retard growth

The data are already arrayed in increasing order. We let Y_1 denote control and Y_2 denote stress. For the K_1 count, we note that there is one Y_2 less than the first Y_1; there are ten Y_2's less than the second Y_1; there are twelve Y_2's less than the third, fourth, fifth, sixth, and seventh Y_1; there are twelve Y_2's less than the eighth Y_1 and one equal to it; and there are thirteen Y_2's less than the ninth through thirteenth Y_1. Thus,

$$K_1 = 1 + 10 + 12 + 12 + 12 + 12 + 12 + 12.5 + 13 + 13 + 13 + 13 + 13 = 148.5.$$

For the K_2 count, we note that there are no Y_1's less than the first Y_2; there is one Y_1 less than the second through tenth Y_2; there are two Y_1's less than the eleventh and twelfth Y_2; and there are seven Y_1's less than the thirteenth Y_2 and one equal to it. Thus,

$$K_2 = 0 + 1 + 1 + 1 + 1 + 1 + 1 + 1 + 1 + 1 + 2 + 2 + 7.5 = 20.5.$$

To check the counts, we verify that

$$K_1 + K_2 = 148.5 + 20.5 = 169 = (13)(13) = (n_1)(n_2).$$

To check the directionality of the data, we note that $K_1 > K_2$, which suggests a tendency for the Y_1's to be larger than the Y_2's, which would indicate that stress retards growth. Thus, the data do deviate from H_0 is the direction specified by H_A.

The Wilcoxon-Mann-Whitney test statistic is the larger of the two counts K_1 and K_2; thus $U_s = 148.5$.

Looking in Table 6 under n = 13 and n' = 13, we find that for a directional alternative, the largest entry is 139, for which the non-directional P-value is 0.0042. Thus, the P-value is bracketed as

   P < 0.0021.

At significance level α = 0.01, we reject H_0, since P < α. There is sufficient evidence (P < 0.0021) to conclude that stress tends to retard growth.

CHAPTER 8
Comparison of Paired Samples

8.2.1 (a) The standard deviation of the four sample differences is given as 0.68. The standard error is

$$SE_{\bar{D}} = \frac{s_D}{\sqrt{n_D}} = \frac{0.68}{\sqrt{4}} = 0.34.$$

(b) H_0: The mean yields of the two varieties are the same ($\mu_1 = \mu_2$)

H_A: The mean yields of the two varieties are different ($\mu_1 \neq \mu_2$)

$t_s = -1.65/0.34 = -4.85$. With df $= 3$, Table 4 gives $t_{0.01} = 4.541$ and $t_{0.005} = 5.841$; thus, $0.01 < P$-value < 0.02. At significance level $\alpha = 0.05$, we reject H_0 if $P < 0.05$. Since $0.01 < P < 0.02$, we reject H_0. There is sufficient evidence ($0.01 < P < 0.02$) to conclude that Variety 2 has a higher mean yield than Variety 1.

(c) H_0: The mean yields of the two varieties are the same ($\mu_1 = \mu_2$)

H_A: The mean yields of the two varieties are different ($\mu_1 \neq \mu_2$)

$$SE_{(\bar{Y}_1 - \bar{Y}_2)} = \sqrt{\frac{1.76^2}{4} + \frac{1.72^2}{4}} = 1.230.$$

$t_s = -1.65/1.230 = -1.34$. With df $= 6$, Table 4 gives $t_{0.20} = .906$ and $t_{0.10} = 1.440$. Thus, $0.20 < P$-value < 0.40 and we do not reject H_0. There is insufficient evidence ($0.20 < P < 0.40$) to conclude that the mean yields of the two varieties are different. (By contrast, the correct test, in part (b), resulted in rejection of H_0.)

8.2.3 Let 1 denote control and let 2 denote progesterone.

H_0: Progesterone has no effect on cAMP ($\mu_1 = \mu_2$)

H_A: Progesterone has some effect on cAMP ($\mu_1 \neq \mu_2$)

The standard error is

$$SE_{\bar{D}} = \frac{s_D}{\sqrt{n_D}} = \frac{0.40}{\sqrt{4}} = 0.20.$$

The test statistic is

$$t_s = \frac{\bar{d}}{SE_{\bar{D}}} = \frac{0.68}{0.20} = 3.4.$$

To bracket the P-value, we consult Table 4 with df $= 4 - 1 = 3$. Table 4 gives $t_{0.025} = 3.182$ and $t_{0.02} = 3.482$. Thus, the P-value is bracketed as

0.04 < P-value < 0.05.

At significance level $\alpha = 0.10$, we reject H_0 if $P < 0.10$. Since $0.04 < P < 0.05$, we reject H_0. There is sufficient evidence ($0.04 < P < 0.05$) to conclude that progesterone decreases cAMP under these conditions.

8.2.6 (a) Let 1 denote treated side and 2 denote control side. The standard error is

$$SE_{\bar{D}} = \frac{s_D}{\sqrt{n_D}} = \frac{1.118}{\sqrt{15}} = 0.2887.$$

The critical value $t_{0.025}$ is found from Student's t distribution with df $= n_D - 1 = 15 - 1 = 14$. From Table 4 we find that $t_{14,0.025} = 2.145$.

The 95% confidence interval is

$$\bar{d} \pm t_{0.025} SE_{\bar{D}}$$

$$0.117 \pm (2.145)(0.2887)$$

$$(-0.50, 0.74) \text{ or } -0.50 < \mu_1 - \mu_2 < 0.74 \text{ °C}.$$

8.4.1 (a) $B_s = 6$. Looking under $n_D = 9$ in Table 7, we see that there is no entry less than or equal to 6. Therefore, $P > 0.20$.

(b) $B_s = 7$. Looking under $n_D = 9$ in Table 7, the P-value is 0.180.

(c) $B_s = 8$. Looking under $n_D = 9$ in Table 7, the P-value is 0.039.

(d) $B_s = 9$. Looking under $n_D = 9$ in Table 7, the P-value is 0.004.

8.4.4 For the sign test, the hypotheses can be stated as

$H_0: p = 0.5$
$H_A: p > 0.5$

where p denotes the probability that the rat in the enriched environment will have the larger cortex. The hypotheses may be stated informally as

H_0: Weight of the cerebral cortex is not affected by environment
H_A: Environmental enrichment increases cortex weight

There were 12 pairs. Of these, there were 10 pairs in which the relative cortex weight was greater for the "enriched" rat than for his "impoverished" littermate; thus $N_+ = 10$ and $N_- = 2$. To check the directionality of the data, we note that

$N_+ > N_-$

Thus, the data so deviate from H_0 in the direction specified by H_A. The value of the test statistic is

B_s = larger of N_+ and N_-
    = 10.

With $n_D = 12$ and $B_s = 10$ the nondirectional P-value is 0.039. We divide this in half to get P = 0.0195, which is less than $\alpha = 0.05$; thus, we reject H_0. There is sufficient evidence (P = 0.0195) to conclude that environmental enrichment increases cortex weight.

8.4.8 P-value $= (2)(0.5^{15}) = 0.000061$.

8.4.11 If it is expected that one treatment "wins" in every pair, then B_S will equal n_D in this study. For this case, the P-value is computed as $(2)(0.5)^{n_D}$. In order for the sample to be large enough to reject H_0 at $\alpha = 0.05$, n_D must satisfy the equation $(2)(0.5)^{n_D} < 0.05$. The smallest value of n_D that satisfies this equation is $n_D = 6$. The P-value will be 0.03125.

8.5.1 (a) P > 0.20

(b) P = 0.078

(c) P = 0.047

(d) P = 0.016

8.5.3 H_0: Hunger rating is not affected by treatment (mCPP vs. placebo)
    H_A: Treatment does affect hunger rating

The absolute values of the differences are 5, 7, 28, 47, 80, 7, 8, and 20.
The ranks of the absolute differences are 1, 2.5, 6, 7, 8, 2.5, 4, and 5.
The signed ranks are –1, 2.5, -6, -7, -8, 2.5, 4, and –5.
Thus, $W_+ = 2.5 + 2.5 + 4 = 9$ and $W_- = 1 + 6 + 7 + 8 + 5 = 27$.
$W_s = 27$ and $n_D = 8$; reading Table 8 we find P-value > 0.20 and H_0 is not rejected. There is insufficient evidence (P > 0.20) to conclude that treatment has an effect.

8.6.4 No. "Accurate" prediction would mean that the individual differences (d's) are small. To judge whether this is the case, one would need the individual values of the d's; using these, one could see whether most of the magnitudes (|d|'s) are small.

8.S.8 The null and alternative hypotheses are

H_0: The average number of species is the same in pools as in riffles ($\mu_1 = \mu_2$)
H_A: The average numbers of species in pools and in riffles differ ($\mu_1 \neq \mu_2$)

The standard error is

$$SE_{\bar{D}} = \frac{s_D}{\sqrt{n_D}} = \frac{1.86}{\sqrt{15}} = 0.48.$$

The test statistic is

$$t_s = \frac{\bar{d}}{SE_{\bar{D}}} = \frac{2.2}{0.48} = 4.58.$$

To bracket the P-value, we consult Table 4 with df = 15 - 1 = 14. Table 4 gives $t_{0.0005} = 4.140$. Thus, the P-value for the nondirectional test is bracketed as

P < 0.001.

At significance level $\alpha = 0.10$, we reject H_0 if P < 0.10. Since P < 0.001, we reject H_0. There is sufficient evidence (P < 0.001) to conclude that the average number of species in pools is greater than in riffles.

8.S.12 The null and alternative hypotheses are

H_0: Caffeine has no effect on RER ($\mu_1 = \mu_2$)
H_A: Caffeine has some effect on RER ($\mu_1 \neq \mu_2$)

We proceed to calculate the differences, the standard error of the mean difference, and the test statistic.

Subject	Placebo	Caffeine	Difference
1	105	96	9
2	119	99	20
3	92	89	3
4	97	95	2
5	96	88	8
6	101	95	6
7	94	88	6
8	95	93	2
9	98	88	10
Mean			7.33
SD			5.59

The standard error is

$$SE_{\bar{D}} = \frac{s_D}{\sqrt{n_D}} = \frac{5.59}{\sqrt{9}} = 1.86.$$

The test statistic is

$$t_s = \frac{\bar{d}}{SE_{\bar{D}}} = \frac{7.33}{1.86} = 3.94.$$

To bracket the P-value, we consult Table 4 with df = 9 - 1 = 8. Table 4 gives $t_{0.005}$ = 3.355 and $t_{0.0005}$ = 5.041. Thus, the P-value for the nondirectional test is bracketed as

$0.001 < P < 0.01$.

At significance level α = .05, we reject H_0 if $P < 0.05$. Since $P < 0.01$, we reject H_0. To determine the directionality of departure from H_0, we note that

$\bar{d} > 0$;   that is, $\bar{y}_1 > \bar{y}_2$.

There is sufficient evidence ($0.001 < P < 0.01$) to conclude that caffeine tends to decrease RER under these conditions.

UNIT II

II.2 (a) The SE for $\overline{Y}_1 - \overline{Y}_2$ is $\sqrt{\dfrac{3.89^2}{27} + \dfrac{1.06^2}{14}} = 0.80$. The test statistics is $t_s = \dfrac{6.59 - 3.96}{0.80} = 3.29$.

Using 32 degrees of freedom, we have $0.001 < P\text{-value} < 0.01$. Since the P-value is less than the α level of 0.02, we reject H_0.

(b) There is strong evidence that the average ecological footprint for the women differs from that for the men. (Indeed, the data suggest that the average for women is greater than the average for men.) The difference in sample averages cannot easily be explained by chance. (*Note:* The *sample* means are statistically significantly different, so we infer that the population means are *different* [not "significantly different"].)

(c) Using 32 degrees of freedom and calculations from part (a) above, a 95% confidence interval is

$6.59 - 3.96 \pm t_{0.025} \times 0.80$

$6.59 - 3.96 \pm 2.042 \times 0.80$

2.63 ± 1.634

$(1.00, 4.46)$ hectares

Since the interval is entirely above 0.7 hectares, the difference is "ecologically important."

II.4 The confidence interval excludes zero, so we would reject H_0; this means that the P-value is less than 0.05. Thus the P-value is less than 0.10, so we would reject H_0.

II.8 (a) Power goes up as n goes up. This is because the SE goes down as n goes up. A larger sample size gives a smaller standard error; hence, more accuracy in estimating the difference in means. Thus, if there is a true difference, we are more likely to detect it when $n = 18$ than when $n = 12$.

(b) These normal curves have a common SD of 1 unit—the distance from the peak of a curve to the point of inflection; or note that ± 3 units covers essentially all of a distribution. The peaks of the normal curves are separated by 1.5 units, so the effect size is $= 1.5$.

II.9 (a) There are three differences ≥ 31 and two differences ≤ -31. Thus $P\text{-value} = 5/28 \approx 0.1563$.

(b) There is no statistically significant evidence that men and women differ with respect to the mean of variable Y. The $P\text{-value} = 0.1536 \not< \alpha = 0.10$.

(c) This test is a directional test. Here we only consider differences ≥ 31. Thus, the directional P-value is $3/28 \approx 0.107$.

II.20 Since the authors are trying to describe the variability of the actual amount of hydrogen cyanide in the seed source and not variability in the sample mean, the ± 8.3 mg/kg seed would refer to the standard deviation and not the standard error.

CHAPTER 9
Categorical Data: One-Sample Distributions

9.1.2 (a) $Pr\{\tilde{P} = 2/7\} = Pr\{no\ mutants\} = Pr\{all\ are\ non\text{-}mutants\} = (1 - 0.37)^3 = 0.250.$

(b) $Pr\{\tilde{P} = 3/7\} = Pr\{1\ mutant\} = {_3}C_1 p^1 (1 - p)^2$, where $p = 0.37$. This is $= (3)(0.37^1)(0.63^2) = 0.441.$

The smallest possible number of mutants is zero, for which $\tilde{P} = 2/7$. Thus, it is not possible that $\tilde{P}$ will equal zero.

9.1.4 We are concerned with the sampling distribution of $\tilde{P}$, which is governed by a binomial distribution. Letting "success" = "responder," we have $p = 0.2$ and $1 - p = 0.8$. The number of trials is $n = 15$.

(a) The event $\tilde{P} = 5/19$ occurs if there are 3 successes in the 15 trials (because $(3+2)/(15+4) = 5/19$). Thus, to find the probability that $\tilde{P} = 5/19$, we can use the binomial formula ${_n}C_j p^j (1 - p)^{n-j}$ with $j = 3$, so $n - j = 12$:

$$Pr\{\tilde{P} = 5/19\} = {_{15}}C_3 p^3 (1 - p)^{12} = (455)(0.2^3)(0.8^{12}) = 0.2501.$$

(b) The event $\tilde{P} = 2/19$ occurs if there are 0 successes in the 15 trials. Thus, to find the probability that $\tilde{P} = 2/19$, we can use the binomial formula with $j = 0$, so $n - j = 15$:

$$Pr\{\tilde{P} = 2/19\} = {_{15}}C_0 p^0 (1 - p)^{15} = (1)(1)(0.8^{15}) = 0.0352.$$

9.1.5 (a) Letting "success" = "infected," we have $p = 0.25$ and $1 - p = 0.75$. The number of trials is $n = 4$. We then use the binomial formula ${_n}C_j p^j (1 - p)^{n-j}$ with $n = 4$ and $p = 0.25$. The values of $\tilde{P}$ correspond to numbers of successes and failures as follows:

$\tilde{P}$	Number of successes (j)	Number of failures (n - j)
2/8	0	4
3/8	1	3
4/8	2	2
5/8	3	1
6/8	4	0

Thus, we find

(i)	$Pr\{\tilde{P} = 2/8\}$	$=$	${_4}C_0 p^0 (1 - p)^4$	$=$	$(1)(1)(0.75^4)$	$= 0.3164$
(ii)	$Pr\{\tilde{P} = 3/8\}$	$=$	${_4}C_1 p^1 (1 - p)^3$	$=$	$(4)(0.25)(0.75^3)$	$= 0.4219$
(iii)	$Pr\{\tilde{P} = 4/8\}$	$=$	${_4}C_2 p^2 (1 - p)^2$	$=$	$(6)(0.25^2)(0.75^2)$	$= 0.2109$
(iv)	$Pr\{\tilde{P} = 5/8\}$	$=$	${_4}C_3 p^3 (1 - p)^1$	$=$	$(4)(0.25^3)(0.75)$	$= 0.0469$
(v)	$Pr\{\tilde{P} = 6/8\}$	$=$	${_4}C_4 p^4 (1 - p)^0$	$=$	$(1)(0.25^4)(1)$	$= 0.0039$

9.1.9 Because $p = 0.25$, the event E occurs if $\tilde{P}$ is within ± 0.05 of 0.25; this happens if there are 3, 4, or 5 successes, as follows:

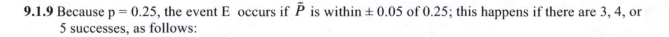

Number of successes (j)	$\tilde{P}$
3	5/24 = 0.208
4	6/24 = 0.250
5	7/24 = 0.292

We can calculate the probabilities of these outcomes using the binomial formula with n = 20 and p = 0.25:

$$\Pr\{\tilde{P} = 0.208\} = {}_{20}C_3 p^3 (1-p)^{17} = (1,140)(0.25^3)(0.75^{17}) = 0.1339$$

$$\Pr\{\tilde{P} = 0.250\} = {}_{20}C_4 p^4 (1-p)^{16} = (4,845)(0.25^4)(0.75^{16}) = 0.1897$$

$$\Pr\{\tilde{P} = 0.292\} = {}_{20}C_5 p^5 (1-p)^{15} = (15,504)(0.25^5)(0.75^{15}) = 0.2023$$

Finally, we calculate $\Pr\{E\}$ by adding these results:
$$\Pr\{E\} = 0.1339 + 0.1897 + 0.2023 = 0.5259.$$

9.2.2 (a) The number of mutants in the sample is y = (100)(0.20) = 20. Thus, $\tilde{p} = (20+2)/(100+4) = 0.212$.

The standard error is

$$SE = \sqrt{\frac{\tilde{p}(1-\tilde{p})}{n+4}} = \sqrt{\frac{0.212(1-0.212)}{100+4}} = 0.040.$$

(b) The number of mutants in the sample is y = (400)(0.20) = 80. Thus, $\tilde{p} = (80+2)/(400+4) = 0.203$.

The standard error is

$$SE = \sqrt{\frac{\tilde{p}(1-\tilde{p})}{n+4}} = \sqrt{\frac{0.203(1-0.203)}{400+4}} = 0.020.$$

9.2.3 (a) The 95% confidence interval is
$$\tilde{p} \pm 1.96\, SE_{\tilde{p}}$$

$$0.212 \pm (1.96)(0.040)$$

$$0.212 \pm 0.078$$

$$(0.134, 0.290) \text{ or } 0.134 < p < 0.290.$$

(b) The 95% confidence interval is
$$\tilde{p} \pm 1.96\, SE_{\tilde{p}}$$

$$0.203 \pm (1.96)(0.020)$$

$$0.203 \pm 0.039$$

$$(0.164, 0.242) \text{ or } 0.164 < p < 0.242.$$

9.2.5 (a) $\tilde{p} = (69+2)/(339+4) = 0.207.$

The standard error is

$$SE = \sqrt{\frac{\tilde{p}(1-\tilde{p})}{n+4}} = \sqrt{\frac{0.207(1-0.207)}{339+4}} = 0.022.$$

The 95% confidence interval is
$$\tilde{p} \pm 1.96\,SE_{\tilde{p}}$$

$$0.207 \pm (1.96)(0.022)$$

$$0.207 \pm 0.043$$

$$(0.164, 0.250) \text{ or } 0.164 < p < 0.250.$$

(b) We are 95% confident that the probability of adverse reaction in infants who receive their first injection of vaccine is between 0.164 and 0.250.

9.2.7 The required n must satisfy the inequality

$$\sqrt{\frac{(\text{Guessed }\tilde{p})(1 - \text{Guessed }\tilde{p})}{n+4}} \leq \text{Desired SE}$$

or

$$\sqrt{\frac{0.6(0.4)}{n+4}} \leq 0.04.$$

It follows that $\dfrac{\sqrt{0.6(0.4)}}{0.04} \leq \sqrt{n+4}$

or $\dfrac{(0.6)(0.4)}{0.04^2} \leq n+4$ or $150 \leq n+4$, so $n \geq 146$.

9.3.4 $\tilde{p} = \dfrac{40+0.5(1.645^2)}{53+1.645^2} = 0.7423$ and $SE = \sqrt{\dfrac{0.7423(1-0.7423)}{53+1.645^2}} = 0.0586$.

90% confidence interval: $0.7423 \pm (1.645)(0.0586)$ or $(0.646, 0.839)$ or $0.646 < p < 0.839$.

Note: In hypothesis testing problems involving the χ^2 statistic, expected frequencies are shown in parentheses.

9.4.1 The hypotheses are

H_0: The model is correct (the population ratio is 12:3:1)
H_A: The model is incorrect (the population ratio is not 12:3:1)

More formally, we can state these as

H_0: Pr{white} = 0.75, Pr{yellow} = 0.1875, Pr{green} = 0.0625
H_A: At least one of the probabilities specified by H_0 is incorrect

We calculate the expected frequencies from H_0 as follows:

White: E =   (.75)(205)   =   153.75
Yellow: E =   (.1875)(205)   =   38.4375
Green: E =   (.0625)(205)   =   12.8125

The observed and expected frequencies (in parentheses) are:

White	Yellow	Green	Total
155 (153.75)	40 (38.4375)	10 (12.8125)	205

The χ^2 test statistic is

$$\chi^2_s = \frac{(155-153.75)^2}{153.75} + \frac{(40-38.4375)^2}{38.4375} + \frac{(10-12.8125)^2}{12.8125} = 0.69.$$

There are 3 categories, so we consult Table 9 with df = 3 - 1 = 2. From Table 9, we find $\chi^2_{2,0.20} = 3.22$. Because $\chi^2_s < \chi^2_{0.20}$, the P-value is bracketed as

P > .20.

At significance level .10, we would reject H_0 if P < 0.10. Since P > 0.20, we do not reject H_0. There is little or no evidence (P > .20) that the model is not correct; the data are consistent with the model.

9.4.2 H_0 and H_A are the same as in Exercise 9.4.1. Because the sample is 10 times as large, the value of χ^2_s is 10 times as large as in Exercise 9.4.1. Thus,

$$\chi^2_s = (10)(0.69) = 6.9.$$

From Table 9, with df = 3 - 1 = 2, we find $\chi^2_{2,0.05} = 5.99$ and $\chi^2_{2,0.02} = 7.82$; thus, the P-value is bracketed as

0.02 < P < 0.05.

At significance level 0.10, we reject H_0 if P < 0.10. Since 0.02 < P < 0.05, we reject H_0. There is sufficient evidence (0.02 < P < 0.05) to conclude that the model is incorrect; the data are not consistent with the model. (Note that, because H_0 is a compound hypothesis, the conclusion for the χ^2 test is nondirectional.)

9.4.8 The hypotheses may be stated informally as

H_0: The drug does not cause tumors
H_A: The drug causes tumors

Consider only the 20 triplets in which at least one tumor occurred. Let T denote the event that a tumor occurs in the treated rat before a tumor occurs in a control rat. If the drug does not cause tumors, then each rat is equally likely to be the first to develop a tumor, so that Pr{T} would be 1/3. On the other hand, if the drug does cause tumors, then the treated rat is at higher risk, so that Pr{T} would be greater than 1/3. Thus, the hypotheses can be stated formally as

$$H_0: Pr\{T\} = \frac{1}{3}$$

$$H_A: Pr\{T\} > \frac{1}{3}$$

Because H_A is directional, we begin by checking the directionality of the data. The estimated probability of T is

$$\hat{Pr}\{T\} = \frac{12}{20} = 0.6.$$

We note that

$$\hat{Pr}\{T\} > \frac{1}{3}.$$

Thus, the data do deviate from H_0 in the direction specified by H_A. We proceed to the calculation of the test statistic. The following are the observed and expected frequencies (in parentheses):

Tumor first in treated rat	Tumor first in control rat
12 (6.67)	8 (13.33)

The expected frequencies are calculated as $\frac{1}{3}(20)$ and $\frac{2}{3}(20)$. The χ^2 statistic is

$$\chi^2_s = \frac{(12-6.67)^2}{6.67} + \frac{(8-13.33)^2}{13.33} = 6.4.$$

There are 2 categories, so we consult Table 9 with df = 2 - 1 = 1. From Table 9, we find $\chi^2_{1,0.02} = 5.41$ and $\chi^2_{1,0.01} = 6.63$, so $\chi^2_{1,0.01} < \chi^2_s < \chi^2_{1,0.02}$. Because H_A is directional, the column headings (0.02 and 0.01) must be cut in half to bracket the P-value; thus P-value is bracketed as

$$0.005 < P < 0.01.$$

At significance level 0.01, we reject H_0 if P < 0.01. Since 0.005 < P < 0.01, we reject H_0. There is sufficient evidence (0.005 < P < 0.01) to conclude that the drug does cause tumors.

9.S.2 (a) For this population, p = 1/5 = 0.20. Letting Y = the number of adults in a random sample of size 16 flatworms we have $\Pr\{\tilde{P}=p\}=\Pr\{\tilde{P}=0.20\}=\Pr\{Y=2\}={}_{16}C_2(0.20)^2(0.80)^{14}=0.2111$.

(b)
$$\Pr\{p-0.05\le\tilde{P}\le p+0.05\}=\Pr\{0.15\le\tilde{P}\le0.25\}=\Pr\{1\le Y\le3\}$$
$$={}_{16}C_1(0.20)^1(0.80)^{15}+{}_{16}C_2(0.20)^2(0.80)^{14}+{}_{16}C_3(0.20)^3(0.80)^{13}$$
$$=0.1126+0.2111+0.2462$$
$$=0.5700$$

9.S.3 $\tilde{p}=(97+2)/(123+4)=0.780$.

The standard error is

$$SE=\sqrt{\frac{\tilde{p}(1-\tilde{p})}{n+4}}=\sqrt{\frac{0.780(1-0.780)}{123+4}}=0.037.$$

The 95% confidence interval is
$$\tilde{p}\pm1.96\,SE_{\tilde{p}}$$

$$0.780\pm(1.96)(0.037)$$

$$0.780\pm0.073$$

$$(0.707,0.853)\ \text{ or }\ 0.707<p<0.853.$$

9.S.14 (a) The hypotheses are

H_0: Directional choice under cloudy skies is random (Pr{toward} = 0.25, Pr{away} = 0.25, Pr{right} = 0.25, Pr{left} = 0.25)
H_A: Directional choice under cloudy skies is not random

The expected frequencies, under H_0, are

Toward: E = (0.25)(50) = 12.5
Away: E = (0.25)(50) = 12.5
Right: E = (0.25)(50) = 12.5
Left: E = (0.25)(50) = 12.5

The observed and expected frequencies (in parentheses) are

Toward	Away	Right	Left	Total
18 (12.5)	12 (12.5)	13 (12.5)	7 (12.5)	50

The χ^2 statistic is
$$\chi^2_s=\frac{(18-12.5)^2}{12.5}+\frac{(12-12.5)^2}{12.5}+\frac{(13-12.5)^2}{12.5}+\frac{(7-12.5)^2}{12.5}=4.88.$$

There are four categories, so we consult Table 9 with df $= 4 - 1 = 3$. From Table 9, we find $\chi^2_{3,0.20} = 4.64$ and $\chi^2_{3,0.10} = 6.25$. Because $\chi^2_{0.20} < \chi^2_{s} < \chi^2_{0.10}$, the P-value is bracketed as

$0.10 < P < 0.20$.

At significance level $\alpha = 0.05$, we reject H_0 if $P < 0.05$. Since $0.10 < P < 0.20$, we do not reject H_0. There is insufficient evidence ($0.10 < P < 0.20$) to conclude that directional choice is not random.

9.S.16 The null and alternative hypotheses are

H_0: The probability of an egg being on a particular type of bean is 0.25 for all four types of beans
H_A: H_0 is false (at least one of the probabilities is not 0.25)

Under H_0, the expected number of eggs for each type of bean is (0.25)(Total), which is $(0.25)(711) = 177.75$. The observed and expected frequencies are

Pinto	Cowpea	Navy	Northern
167 (177.75)	176 (177.75)	174 (177.75)	194 (177.75)

The test statistic is

$$\chi^2_s = \frac{(167-177.75)^2}{177.75} + \frac{(176-177.75)^2}{177.75} + \frac{(174-177.75)^2}{177.75} + \frac{(194-177.75)^2}{177.75} = 2.23.$$

There are 4 categories, so df $= 4 - 1 = 3$. Table 9 gives $\chi^2_{3,0.20} = 4.64$, so $P > 0.20$ and we do not reject H_0. There is insufficient evidence ($P > 0.20$) to conclude that cowpea weevils prefer one type of bean over the others.

CHAPTER 10
Categorical Data: Relationships

10.2.3 (a) To have $\chi^2_s = 0$, the columns of the table (and the rows of the table) must be proportional to each other, as in the following table:

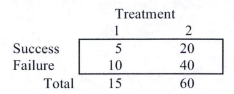

	Treatment	
	1	2
Success	5	20
Failure	10	40
Total	15	60

(b) The estimated probabilities of success are $\hat{p}_1 = 5/15 = 1/3$ and $\hat{p}_2 = 20/60 = 1/3$. Yes, these proportions are equal.

10.2.5 The hypotheses are

> H_0: Mites do not induce resistance to wilt
> H_A: Mites do induce resistance to wilt

Letting p denote the probability of wilt and letting 1 denote mites and 2 denote no mites, the hypotheses may be stated as

> H_0: $p_1 = p_2$
> H_A: $p_1 < p_2$

Because H_A is directional, we begin by checking the directionality of the data. The estimated probabilities of wilt disease are

$$\hat{p}_1 = \frac{11}{26} \approx 0.42;$$

$$\hat{p}_2 = \frac{17}{21} \approx 0.81.$$

We note that

$$\hat{p}_1 < \hat{p}_2.$$

Thus, the data do deviate from H_0 in the direction specified by H_A. We proceed to the calculation of the test statistic. The expected frequency for any given cell is found from the formula

$$E = \frac{(\text{Row total}) \times (\text{Column total})}{\text{Grand total}}$$

The following table shows the observed and expected frequencies (in parentheses):

	Mites	No mites	Total

Wilt disease	11 (15.49)	17 (12.51)	28
No wilt disease	15 (10.51)	4 (8.49)	19
Total	26	21	47

The χ^2 test statistic is

$$\chi_s^2 = \frac{(11-15.49)^2}{15.49} + \frac{(17-12.51)^2}{12.51} + \frac{(15-10.51)^2}{10.51} + \frac{(4-8.49)^2}{8.49} = 7.21.$$

We consult Table 9 with df $= 1$. From Table 9, we find $\chi_{1,0.01}^2 = 6.63$ and $\chi_{1,0.001}^2 = 10.83$, so $\chi_{1,0.001}^2 < \chi_s^2 < \chi_{1,0.01}^2$. Because H_A is directional, the column headings (0.01 and 0.001) must be cut in half to bracket the P-value; thus P-value is bracketed as

$0.0005 < P < 0.005.$

At significance level 0.01, we reject H_0 if $P < 0.01$. Since $0.0005 < P < 0.005$, we reject H_0. There is sufficient evidence ($0.0005 < P < 0.005$) to conclude that mites do induce resistance to wilt.

10.2.10 Let p denote the probability of response, let 1 denote simultaneous, and let 2 denote sequential administration. The hypotheses are

H_0: The two timings are equally effective ($p_1 = p_2$)
H_A: The two timings are not equally effective ($p_1 \neq p_2$)

The expected frequency for any given cell is found from the formula

$$E = \frac{(\text{Row total}) \times (\text{Column total})}{\text{Grand total}}$$

The following table shows the observed and expected frequencies (in parentheses):

	Simultaneous	Sequential	Total
Response	11 (8.30)	3 (5.70)	14
No response	5 (7.70)	8 (5.30)	13
Total	16	11	27

The χ^2 test statistic is

$$\chi_s^2 = \frac{(11-8.3)^2}{8.3} + \frac{(3-5.7)^2}{5.7} + \frac{(5-7.7)^2}{7.7} + \frac{(8-5.3)^2}{5.3} = 4.48.$$

We consult Table 9 with df $= 1$. From Table 9, we find $\chi_{1,0.05}^2 = 3.84$ and $\chi_{1,0.02}^2 = 5.41$, so $\chi_{1,0.02}^2 < \chi_s^2 < \chi_{1,0.05}^2$. Thus P-value is bracketed as

$0.02 < P < 0.05.$

At significance level 0.05, we reject H_0 if $P < 0.05$. Since $0.02 < P < 0.05$, we reject H_0. To determine directionality, we calculate

$$\hat{p}_1 = \frac{11}{16} \approx 0.69,$$

$$\hat{p}_2 = \frac{3}{11} \approx 0.27,$$

and we note that

$$\hat{p}_1 > \hat{p}_2.$$

There is sufficient evidence ($0.02 < P < 0.05$) to conclude that the simultaneous timing is superior to the sequential timing.

10.2.13 Let p denote the probability of hemorrhaging, let 1 denote the ancrod group, and let 2 denote the placebo group.

$H_0: p_1 = p_2$
$H_A: p_1 \neq p_2$

	Ancrod	Placebo	Total
Yes	13 (8.928)	5 (9.072)	18
No	235 (239.072)	247 (242.928)	482
Total	248	252	500

$\chi^2_s = 3.82$. With df = 1, Table 9 gives $\chi^2_{0.10} = 2.71$ and $\chi^2_{0.05} = 3.84$, so $0.05 < P < 0.10$. We do not reject H_0; there is insufficient evidence ($0.05 < P < 0.10$) to conclude that hemorrhaging is more likely under one treatment than under the other.

10.3.3 (a) The estimated conditional probability that a patient dies, given that he had surgery, is

$$\hat{Pr}\{D|S\} = \frac{83}{347} = 0.239. \quad \text{Likewise,} \quad \hat{Pr}\{D|WW\} = \frac{106}{348} = 0.3046.$$

The estimated conditional probability that a patient had surgery, given that the patient died, is

$$\hat{Pr}\{S|D\} = \frac{83}{189} = 0.7615. \quad \text{Likewise,} \quad \hat{Pr}\{S|A\} = \frac{264}{506} = 0.5217.$$

(b) H_0: There is no relationship between treatment and survival.
  H_A: There is a relationship between treatment and survival.

$\chi^2_s = 3.75$. With df = 1, Table 9 gives $\chi^2_{0.10} = 2.71$ and $\chi^2_{0.05} = 3.84$, so $0.05 < P < 0.10$.

We do not reject H_0; there is no significant evidence ($0.05 < P < 0.10$) that treatment and survival are related.

10.3.4 The following table shows the data arranged as a contingency table.

		Preferred Hand		
		Right	Left	Total
Preferred	Right	2012	121	2133
Foot	Left	142	116	258
	Total	2154	237	2391

Let RH, LH, RF, and LF denote right-handed, left-handed, right-footed, and left-footed.

(a) The estimated conditional probability that a woman is right-footed, given that she is right-handed, is

$$\hat{Pr}\{RF|RH\} = \frac{2012}{2154} = 0.934.$$

(b) The estimated conditional probability that a woman is right-footed, given that she is left-handed, is

$$\hat{Pr}\{RF|LH\} = \frac{121}{237} = 0.511.$$

(c) We test independence of hand preference and foot preference using a χ^2 test. We calculated expected frequencies from the formula

$$E = \frac{(\text{Row total})\times(\text{Column total})}{\text{Grand total}}$$

The following table shows the expected frequencies:

		Preferred Hand		
		Right	Left	Total
Preferred	Right	1921.57	211.43	2133
Foot	Left	232.43	25.57	258
	Total	2154	237	2391

The χ^2 test statistic is

$$\chi^2_s = \frac{(2012-1921.57)^2}{1921.57} + \frac{(121-211.43)^2}{211.43} + \frac{(142-232.43)^2}{232.43} + \frac{(116-25.57)^2}{25.57} = 398.$$

(d) The null hypothesis can be expressed as

$$H_0: Pr\{RF|RH\} = 0.5 = Pr\{LF|RH\}$$

This requires a goodness-of-fit test. The expected frequencies are calculated from H_0 as

Right-footed: E = (0.5)(2154) = 1077
Left-footed: E = (0.5)(2154) = 1077

The observed frequencies are

Right-footed: 2012
Left-footed: 142

The χ^2 test statistic is

$$\chi^2_s = \frac{(2012-1077)^2}{1077} + \frac{(142-1077)^2}{1077} = 1{,}623.$$

10.4.1 Tables that more strongly support H_A are those with fewer than 2 deaths on treatment B. There are two such tables:

5	1
9	15

6	0
8	16

10.5.3 (a) Letting UP denote ulcer patient and C denote control, the hypotheses are

H_0: The blood type distributions are the same for ulcer patients and controls ($\Pr\{O|UP\} = \Pr\{O|C\}$, $\Pr\{A|UP\} = \Pr\{A|C\}$, $\Pr\{B|UP\} = \Pr\{B|C\}$, $\Pr\{AB|UP\} = \Pr\{AB|C\}$)

H_A: The blood type distributions are not the same

The test statistic is $\chi^2_s = 49.0$. The degrees of freedom are df $= (4 - 1)(2 - 1) = 3$. From Table 9 we find that $\chi^2_{3,0.0001} = 21.11$. Because $\chi^2_s < \chi^2_{0.0001}$, the P-value is bracketed as

$P < 0.0001.$

At significance level $\alpha = 0.01$, we will reject H_0 if $P < 0.01$. Since $P < 0.0001$, we reject H_0 and conclude that the blood type distribution of ulcer patients is different from that of controls.

10.5.5 (a) The hypotheses are

H_0: Change in ADAS-Cog score is independent of treatment

H_A: Change in ADAS-Cog score is related to treatment

The test statistic is $\chi^2_s = 10.26$. The degrees of freedom are df $= (2 - 1)(5 - 1) = 4$. From Table 9 we find that $\chi^2_{4,0.05} = 9.49$ and $\chi^2_{4,0.02} = 11.67$. Thus, $0.02 < P < 0.05$, so we reject H_0. There is sufficient evidence ($0.02 < P < 0.05$) to conclude that EGb and placebo are not equally effective.

10.6.2 This analysis is not appropriate because the observational units (mice) are nested within the units (litters) that were randomly allocated to treatments. This hierarchical structure casts doubt on the condition that the observations on the 224 mice are independent, especially in light of the investigator's comment that the response varied considerably from litter to litter.

10.7.3 $\tilde{p}_1 = 33/107 = 0.3084$, $\tilde{p}_2 = 21/109 = 0.1927$

$$SE_{(\tilde{p}_1 - \tilde{p}_2)} = \sqrt{\frac{(0.3084)(0.6916)}{107} + \frac{(0.1927)(0.8073)}{109}} = 0.0585.$$

$(0.3084 - 0.1927) \pm (1.96)(0.0585)$

$(0.001, 0.230)$ or $0.001 < p_1 - p_2 < 0.230$. No; the confidence interval suggests that bed rest may actually be harmful.

10.7.5 (a) $\tilde{p}_1 = 912/1657 = 0.5504$, $\tilde{p}_2 = 4579/10002 = 0.4578$.

$$SE_{(\tilde{p}_1 - \tilde{p}_2)} = \sqrt{\frac{(0.5504)(0.4496)}{1657} + \frac{(0.4578)(0.5422)}{10002}} = 0.0132.$$

$(0.5504 - 0.4578) \pm (1.96)(0.0132)$

$(0.067, 0.118)$ or $0.067 < p_1 - p_2 < 0.118$.

(b) We are 95% confident that the proportion of persons with type O blood among ulcer patients is higher than the proportion of persons with type O blood among healthy individuals by between 0.067 and 0.118. That is, we are 95% confident that p_1 exceeds p_2 by between 0.067 and 0.118.

10.8.1 The data are

		Case	
		No	Yes
Control	No	107	30
	Yes	13	5

Note: There is an error in the first printing of the 2nd edition, whereby the headings of "Yes" and "No" were reversed.

The hypotheses are

H_0: There is no association between oral contraceptive use and stroke ($p = 0.5$)
H_A: There is an association between oral contraceptive use and stroke ($p \neq 0.5$)

where p denotes the probability that a discordant pair will be Yes(case)/No(control).

The test statistic for McNemar's test is $\chi^2_s = \dfrac{(13-30)^2}{13+30} = 6.72$.

Looking in Table 9, with df = 1, we see that $\chi^2_{0.01} = 6.63$ and $\chi^2_{0.001} = 10.83$. Thus, $0.001 < P < 0.01$, so we reject H_0. There is sufficient evidence ($0.001 < P < 0.01$) to conclude that stroke victims are more likely to be oral contraceptive users ($p > 0.5$).

10.9.1 (a) (i) $\hat{p}_1 = 25/517 = 0.04836$ and $\hat{p}_2 = 23/637 = 0.0361$. The relative risk is $\hat{p}_1 / \hat{p}_2 = 0.04836/0.03611 = 1.339$.

(ii) The odds ratio is $\dfrac{(25)(614)}{(23)(492)} = 1.356$.

(b) (i) $\hat{p}_1 = 12/105 = 0.11429$ and $\hat{p}_2 = 8/92 = 0.08696$. The relative risk is $\hat{p}_1 / \hat{p}_2 = 0.11429/0.08696 = 1.314$.

(ii) The odds ratio is $\dfrac{(12)(84)}{(8)(93)} = 1.355$.

10.9.7 (a) The sample odds ratio is $\dfrac{(309)(1341)}{(266)(1255)} = 1.2413$.

(b) $\ln(\hat{\theta}) = 0.2161$; $SE_{\ln(\hat{\theta})} = \sqrt{\dfrac{1}{309} + \dfrac{1}{266} + \dfrac{1}{1255} + \dfrac{1}{1341}} = 0.0924$. The 95% confidence interval for $\ln(\theta)$ is $0.2161 \pm (1.96)(0.0924)$, which is $(0.0350, 0.3972)$. $e^{0.0350} = 1.036$ and $e^{0.3972} = 1.488$. The 95% confidence interval for θ is $(1.036, 1.488)$.

(c) We are 95% confident that taking heparin increases the odds of a negative response by a factor of between 1.036 and 1.488 when compared to taking enoxaparin. Since a negative outcome is fairly rare, we can say that we are 95% confident that the probability of a negative outcome is between 1.036 and 1.488 times higher for patients given heparin than for patients given enoxaparin.

10.S.3 Let p denote the probability of female and let 1 and 2 denote warm and cold environments.

(a) H_0: The sex ratio is 1:1 in the warm environment ($p_1 = 0.5$)

H_A: Sex ratio is not 1:1 in the warm environment ($p_1 \neq 0.5$)

The expected frequencies are calculated from H_0 as follows:

Female: $E = (0.5)(141) = 70.5$
Male: $E = (0.5)(141) = 70.5$

The observed and expected frequencies (in parentheses) are

Female	Male	Total
73 (70.5)	68 (70.5)	141

The χ^2 statistic is

$$\chi^2_s = \frac{(73 - 70.5)^2}{70.5} + \frac{(68 - 70.5)^2}{70.5} = 0.18.$$

There are two categories (female and male), so we consult Table 9 with df = 2 - 1 = 1. From Table 9, we find $\chi^2_{1,0.20} = 1.64$. Because $\chi^2_s < \chi^2_{0.20}$, the P-value is bracketed as

$P > 0.20$.

At significance level $\alpha = 0.05$, we reject H_0 if $P < 0.05$. Since $P > 0.20$, we do not reject H_0. There is insufficient evidence ($P > 0.20$) to conclude that the sex ratio is not 1:1 in the warm environment.

(c) The hypotheses are

H_0: Sex ratio is the same in the two environments ($p_1 = p_2$)
H_A: Sex ratio is not the same in the two environments ($p_1 \neq p_2$)

We calculate the expected frequencies under H_0 from the formula

$$E = \frac{(\text{Row total}) \times (\text{Column total})}{\text{Grand total}}$$

The following table shows the observed and expected frequencies (in parentheses):

	Environment		
	Warm	Cold	Total
Male	68 (59.13)	62 (70.87)	130
Female	73 (81.87)	107 (98.13)	180
Total	141	169	310

The χ^2 test statistic is

$$\chi_s^2 = \frac{(68-59.13)^2}{59.13} + \frac{(62-70.87)^2}{70.87} + \frac{(73-81.87)^2}{81.87} + \frac{(107-98.13)^2}{98.13} = 4.20.$$

The degrees of freedom are df $= (2-1)(2-1) = 1$. Table 9 shows that $\chi_{1,0.05}^2 = 3.84$ and $\chi_{1,0.02}^2 = 5.41$. Thus, $0.02 < P < 0.05$, so H_0 is rejected. To determine directionality, we calculate

$$\hat{p}_1 = \frac{73}{141} = 0.52,$$

$$\hat{p}_2 = \frac{107}{169} = 0.63,$$

and we note that $\hat{p}_1 < \hat{p}_2$.

There is sufficient evidence ($0.02 < P < 0.05$) to conclude that the probability of a female is higher in the cold than the warm environment.

10.S.12 The null and alternative hypotheses are

H_0: Site of capture and site of recapture are independent ($\Pr\{RI|CI\} = \Pr\{RI|CII\}$)
H_A: Flies preferentially return to their site of capture ($\Pr\{RI|CI\} > \Pr\{RI|CII\}$)

where C and R denote capture and recapture and I and II denote the sites.

Because H_A is directional, we begin by checking the directionality of the data. We calculate

$$\hat{\Pr}\{RI|CI\} = \frac{78}{134} = 0.58,$$

$$\hat{\Pr}\{RI|CII\} = \frac{33}{91} = 0.36,$$

and we note that $\hat{\Pr}\{RI|CI\} > \hat{\Pr}\{RI|CII\}$.

Thus, the data deviate from H_0 in the direction specified by H_A.

The test statistic is $\chi^2_s = 10.44$. From Table 9 with df $= 1$, we find $\chi^2_{1,0.01} = 6.63$ and $\chi^2_{1,0.001} = 10.83$. We cut the column headings in half for the directional test. Thus, $0.0005 < P < 0.005$ and we reject H_0. At the 0.01 level, there is sufficient evidence ($0.0005 < P < 0.005$) to conclude that flies preferentially return to their site of capture.

10.S.14 (a) The sample odds ratio is $\dfrac{(25339)(686)}{(8914)(1141)} = 1.709.$

(b) $\ln\left(\hat{\theta}\right) = .5359$; $SE_{\ln(\hat{\theta})} = \sqrt{\dfrac{1}{25339} + \dfrac{1}{1141} + \dfrac{1}{8914} + \dfrac{1}{686}} = 0.0499.$ The 95% confidence interval for $\ln(\theta)$ is $0.5359 \pm (1.96)(0.0499)$, which is $(0.438, 0.634)$. $e^{0.438} = 1.55$ and $e^{0.634} = 1.89$. The 95% confidence interval for θ is $(1.55, 1.89)$ or $1.55 < \theta < 1.89$.

(c) The odds ratio gives the (estimated) odds of survival for men compared to women. Another way to say this is that it gives the odds of death for women compared to men. This ratio (of 1.709) is a good approximation to the relative risk of death for women compared to men (which is 1.658), because death is fairly rare.

UNIT III

III.2 (a) The two sample proportions are $5/16 = 0.3125$ and $18/30 = 0.60$; thus, the data support H_A. We have $0.025 < P\text{-value} < 0.05$, so we reject H_0. There is evidence that men are less likely than women to be involved in community service.

(b) We can focus on the upper left cell, which contains a 5. More extreme tables would have 4, 3, 2, 1, or 0 here, so there are 6 tables to consider.

III.3 (a) If germination rate is independent of type of seed, then the best estimate of germination rate is $32/57$. Applying that rate to 20 Okra seeds gives $20 \times 32/57 = 11.23$ expected to germinate.

(b) $\chi^2 = 1.84$, df $= 2$. P $> 0.20 > 0.05$, so we retain H_0.

III.5 (a) If $\tilde{p} = 0.25$ then we want $\sqrt{\dfrac{0.25 * 0.75}{n+4}} \leq 0.06$ so $n + 4 \geq \left(\dfrac{\sqrt{0.25 * 0.75}}{0.06}\right)^2$ which equals 52.08.

Thus, we want $n \geq 48.08$, so we must take $n = 49$.

(b) If we don't have a guess for $\tilde{p}$ then we use $\tilde{p} = 0.50$. This gives $\sqrt{\dfrac{0.5 * 0.5}{n+4}} \leq 0.06$ so

$n + 4 \geq \left(\dfrac{\sqrt{0.5 * 0.5}}{0.06}\right)^2$ which equals 69.44. Thus we want $n \geq 65.44$, so we must take $n = 66$.

III.10 (a) True. The standard error of the estimated proportion is largest when $p = 0.50$.

(b) False. A goodness-of-fit test can be conducted for any null hypothesis that specifies probabilities for each of the possible categories of a response variable, but there is no need for those probabilities to be equal.

(c) False. Either a goodness-of-fit test or a test of independence can be conducted for observational data or for experimental data.

(d) True. If the observed data perfectly agree with the expected values from the null hypothesis, then each term in the chi-square calculation will be zero, so the test statistic will be zero.

CHAPTER 11
Comparing the Means of Many Independent Samples

11.2.1 We have n. = 4 + 3 + 4 = 11;

$$\sum_{i=1}^{I}\sum_{j=1}^{n_i} y_{ij} = 48 + 39 + 42 + 43 + 40 + 48 + 44 + 39 + 30 + 32 + 35 = 440;$$

$$\bar{\bar{y}} = \frac{440}{11} = 40.$$

(a) SS(between) = $(4)(43-40)^2 + (3)(44-40)^2 + (4)(34-40)^2 = 228$;

SS(within) = $(48-43)^2 + (39-43)^2 + (42-43)^2 + (43-43)^2$
$\qquad + (40-44)^2 + (48-44)^2 + (44-44)^2$
$\qquad + (39-34)^2 + (30-34)^2 + (32-34)^2 + (35-34)^2 = 120.$

(b) SS(total) = $(48-40)^2 + (39-40)^2 + (42-40)^2 + (43-40)^2$
$\qquad + (40-40)^2 + (48-40)^2 + (44-40)^2$
$\qquad + (39-40)^2 + (30-40)^2 + (32-40)^2 + (35-40)^2 = 348.$

Verification: SS(between) + SS(within) = SS(total);
$$228 + 120 = 348.$$

(c) df(between) = I - 1 = 3 - 1 = 2; MS(between) = $\dfrac{\text{SS(between)}}{\text{df(between)}} = \dfrac{228}{2} = 114$;

df(within) = n. - I = 11 - 3 = 8; MS(within) = $\dfrac{\text{SS(within)}}{\text{df(within)}} = \dfrac{120}{8} = 15$;

$$s_{pooled} = \sqrt{\text{MS(within)}} = \sqrt{15} = 3.87.$$

11.2.4 (a) We find SS(between) by subtraction: SS(between) = 472 - 337 = 135.

We find df(total) by adding df(between) and df(within): df(total) = 3 + 12 = 15.

We find MS(within) by division: MS(within) = SS(within)/df(within) = 337/12 = 28.08.

The completed table is

Source	df	SS	MS
Between groups	3	135	45
Within groups	12	337	28.08
Total	15	472	

(b) We have df(between) = 3 = I - 1, so I = 4.

(c) We have df(total) = 15 = n. - 1, so n. = 16.

11.4.2 (a) The hypotheses are

H_0: The stress conditions all produce the same mean lymphocyte concentration
$(\mu_1 = \mu_2 = \mu_3 = \mu_4)$
H_A: Some of the stress conditions produce different mean lymphocyte concentrations
(the μ's are not all equal)

The number of groups is I = 4 and the total number of observations is n.= 48. Thus, df(between) = I - 1 = 3 and df(within) = n.- I = 44.

Source	df	SS	MS
Between groups	3	89.036	29.68
Within groups	44	340.24	7.733
Total	47	429.28	

The test statistic is $F_s = \dfrac{MS(between)}{MS(within)} = \dfrac{29.68}{7.733} = 3.84$. With df = 3 and 40 (the closest value to 44), Table 10 gives $F_{0.02} = 3.67$ and $F_{0.01} = 4.31$. Thus, we have $0.01 < P < 0.02$. Since $P < \alpha$, we reject H_0. There is sufficient evidence $(0.01 < P < 0.02)$ to conclude that some of the stress conditions produce different mean lymphocyte concentrations.

(b) $s_{pooled} = \sqrt{MS(within)} = \sqrt{7.733} = 2.78$ cells/ml x 10^{-6}.

11.4.3 (a) The null hypothesis is

H_0: Mean HBE is the same in all three populations

(d) $s_{pooled} = \sqrt{MS(within)} = \sqrt{208.7} = 14.4$ pg/ml.

11.6.2 There is no single correct answer to this exercise, because it involves randomization. Within each litter (block), one animal will be allocated to each treatment. The random allocation is carried out separately for each litter.

To illustrate the procedure, we allocate the animals in litter 1. We assign the animals identification numbers 1, 2, 3, 4, 5. We then read random digits from a calculator or from Table 1. For instance, suppose the random digits are

2, 0, 1, 9, 4, 1, 7, 8, 2, 5, ...

Then we would assign animal 2 to treatment 1, animal 1 to treatment 2, animal 4 to treatment 3, animal 5 to treatment 4, and the remaining animal (3) to treatment 5. (We ignore the digits 0, 9, 7, and 8 because no animals have these identification numbers. Likewise, we ignore the second 1 in the list because animal 1 has already been assigned to a treatment by the time we encounter this number.)

After proceeding similarly (using new random digits) for each litter, we might obtain the following allocation:

Treatment	Piglet				
	Litter 1	Litter 2	Litter 3	Litter 4	Litter 5
1	2	5	2	4	5
2	1	4	1	1	2
3	4	2	5	2	4
4	5	3	3	3	3
5	3	1	4	5	1

11.6.5 Plan II is better. We want units within a block to be similar to each other; plan II achieves this. Under plan I the effect of rain would be confounded with the effect of a variety.

11.7.2 (a)

Source	df	SS	MS
Between species	1	2.19781	2.19781
Between flooding levels	1	2.25751	2.25751
Interaction	1	0.097656	0.097656
Within groups	12	0.47438	0.03953
Total	15	5.027356	

(b) There are 1 numerator and 12 denominator degrees of freedom.

(c) We do not reject H_0. There is insufficient evidence (P = 0.142) to conclude that there is an interaction present.

(d) $F_s = 2.19781/0.03953 = 55.60$.

(e) H_0 is rejected. There is strong evidence (P = 0.000008) to conclude that species affects ATP concentration.

(f) $s_{pooled} = \sqrt{0.03953} = 0.199$.

11.7.4 $F_s = \dfrac{31.33/1}{30648.81/(223-4)} = 31.33/139.95 = 0.22$. With df = 1 and 140, Table 10 gives $F_{0.20} = 1.66$.

Thus, P > 0.20 and we do not reject H_0. There is insufficient evidence (P > 0.20) to conclude that there is an interaction present.

11.8.2 (a) For women with no children, the age-adjusted blood pressure is the following linear combination:

$L = (0.17)(113) + (0.29)(118) + (0.31)(125) + (0.23)(134) = 123$ mm Hg.

(b) For women with five or more children, the age-adjusted blood pressure is the following linear combination:

$$L = (0.17)(114) + (0.29)(116) + (0.31)(124) + (0.23)(138) = 123.2 \text{ mm Hg.}$$

(d) For the linear combination in part (a), the multipliers are

$$0.17, 0.29, 0.31, 0.23.$$

The standard error is

$$SE_L = 18\sqrt{\frac{0.17^2}{230} + \frac{0.29^2}{110} + \frac{0.31^2}{105} + \frac{0.23^2}{123}} = 0.851 \text{ mm Hg.}$$

<u>Note</u>: For exercises that require definition of contrasts, the answer given is not the only correct answer. For example, reversing the sign of all contrasts is also correct.

11.8.7 $L = \dfrac{1}{2}(3.09 - 2.16) + \dfrac{1}{2}(4.48 - 3.26) = 1.075.$

The m's are $\dfrac{-1}{2}, \dfrac{1}{2}, \dfrac{-1}{2}, \dfrac{1}{2}$; $\Sigma m^2 = 1.$

$$SE_L = \sqrt{\frac{0.3481}{9}(1)} = 0.1967.$$

$$1.075 \pm (2.402)(0.1967) \qquad\qquad (df = 30)$$

$$0.60 < \frac{1}{2}(\mu_2 - \mu_1) + \frac{1}{2}(\mu_4 - \mu_3) < 1.55 \text{ gm}$$

or

$$0.60 < \mu_E - \mu_S < 1.55 \text{ gm}$$

where

$$\mu_E = \frac{1}{2}(\mu_{E,Low} + \mu_{E,High}) \text{ and } \mu_S = \frac{1}{2}(\mu_{S,Low} + \mu_{S,High}).$$

11.8.8 (b) Using the means given in Exercise 11.4.1, the value of L is

$$L = 9.81 - \frac{1}{2}(6.28 + 5.97) = 3.685 \text{ nmol/10}^8 \text{ platelets/hour.}$$

To find the standard error of L, we need to calculate s_{pooled}. From Exercise 11.4.1, SS(within) = 418.25 and df(within) = (18 + 16 + 8) - 3 = 39, so

$$s_{pooled} = \sqrt{\frac{418.25}{39}} = \sqrt{10.72}\ .$$

The multipliers in L are 1, $\frac{-1}{2}$, and $\frac{-1}{2}$. The standard error of L is

$$SE_L = \sqrt{10.72\left(\frac{1}{18} + \frac{(-1/2)^2}{16} + \frac{(-1/2)^2}{8}\right)} = 1.048 \text{ nmol}/10^8 \text{ platelets/hour.}$$

11.9.1 For these data, $s_{pooled} = \sqrt{0.2246} = 0.4739$, $SE_{D_{ab}} = 0.4739 \times \sqrt{(1/9) + (1/9)} = 0.2234$, and $t_{40,0.025} \times SE_{D_{ab}} = 2.021 \times 0.2234 = 0.4515$.

The 95% Fisher LSD Intervals are summarized as follows. Intervals in italics do not contain zero and thus indicate statistically significant differences.

Comparison	d_{ab}	$t_{40,0.025} \times SE_{D_{ab}}$	lower 95%	upper 95%
A - B	-0.39	0.4515	-0.842	0.062
A - C	*0.67*	*0.4515*	*0.219*	*1.122*
A - D	*-1.04*	*0.4515*	*-1.492*	*-0.589*
A - E	*-1.01*	*0.4515*	*-1.462*	*-0.559*
B - C	*1.06*	*0.4515*	*0.609*	*1.512*
B - D	*-0.65*	*0.4515*	*-1.102*	*-0.199*
B - E	*-0.62*	*0.4515*	*-1.072*	*-0.169*
C - D	*-1.71*	*0.4515*	*-2.162*	*-1.259*
C - E	*-1.68*	*0.4515*	*-2.132*	*-1.229*
D - E	0.03	0.4515	-0.422	0.482

To display our results as in Table 11.9.4 we have

Treatment	
A	4.37^b
B	4.76^b
C	3.70^a
D	5.41^c
E	5.38^c

Groups sharing a common superscript have means that are not statistically significantly different based on Fisher comparisons with $\alpha_{cw} = 0.05$

11.9.2 For these data, $s_{pooled} = \sqrt{0.2246} = 0.4739$, $SE_{D_{ab}} = 0.4739 \times \sqrt{(1/9)+(1/9)} = 0.2234$, and, using Table 11 to find the Bonferroni adjusted t-multiplier for df = 40 and k = 10, $t_{40,0.0025} \times SE_{D_{ab}} = 2.971 \times 0.2234 = 0.6637$.

The experimentwise 95% (99.5% comparisonwise) Bonferroni Intervals are summarized as follows. Intervals in italics do not contain zero and thus indicate statistically significant differences.

Comparison	d_{ab}	$t_{40,0.025} \times SE_{D_{ab}}$	lower 95%	upper 95%
A - B	-0.39	0.6637	-1.054	0.274
A - C	*0.67*	*0.6637*	*0.006*	*1.334*
A - D	*-1.04*	*0.6637*	*-1.704*	*-0.376*
A - E	*-1.01*	*0.6637*	*-1.674*	*-0.346*
B - C	*1.06*	*0.6637*	*0.396*	*1.724*
B - D	-0.65	0.6637	-1.314	0.014
B - E	-0.62	0.6637	-1.284	0.044
C - D	*-1.71*	*0.6637*	*-2.374*	*-1.046*
C - E	*-1.68*	*0.6637*	*-2.344*	*-1.016*
D - E	0.03	0.6637	-0.634	0.694

To display our results as in Table 11.9.4 we have

Treatment	
A	4.37^{b}
B	4.76^{bc}
C	3.70^{a}
D	5.41^{c}
E	5.38^{c}

Groups sharing a common superscript have means that are not statistically significantly different based on Bonferroni comparisons with $\alpha_{ew} = 0.05$

11.9.4 (a) There is evidence that the mean selenium levels for cows fed diets B, C, or D, each differ from the mean selenium level for cows fed diet A as all three 95% Tukey confidence intervals (B-A, C-A, and D-A) exclude zero.

11.S.1 (a) H_0: The three classes produce the same mean change in fat-free mass ($\mu_1 = \mu_2 = \mu_3$); H_A: At least one class produces a different mean (the μ's are not all equal).

(b)

Source	df	SS	MS
Between groups	2	2.465	1.2325
Within groups	26	50.133	1.9282
Total	28	52.598	

$F_s = 1.2325/1.9282 = 0.64$. We do not reject H_0. There is insufficient evidence (P > 0.20) to conclude that the population means differ.

11.S.3 The hypotheses are

H_0: The mean refractive error is the same in the four populations ($\mu_1 = \mu_2 = \mu_3 = \mu_4$)
H_A: Some of the populations have different mean refractive errors (the μ's are not all equal)

We have I = 4 and n. = 211. Thus,

df(between) = I - 1 = 4 - 1 = 3;

df(within) = n. - I = 211 - 4 = 207.

The ANOVA table is

Source	df	SS	MS
Between groups	3	129.49	43.16
Within groups	207	2506.8	12.11
Total	210	2636.3	

The test statistic is $F_s = \dfrac{MS(between)}{MS(within)} = \dfrac{43.26}{12.11} = 3.56$. With df = 3 and ∞, Table 10 gives $F_{0.02} = 3.28$ and $F_{0.01} = 3.78$. Thus, we have

0.01 < P < 0.02.
Since P < α, we reject H_0. There is sufficient evidence (0.01 < P < 0.02) to conclude that some of the populations have different mean refractive errors.

11.S.13 Let 1, 2, 3, and 4 denote placebo; probucol; multivitamins; and probucol and multivitamins.

(a) $\bar{y}_2 - \bar{y}_1 = 1.79 - 1.43 = 0.36$.

(b) $\bar{y}_4 - \bar{y}_3 = 1.54 - 1.40 = 0.14$.

(c) The contrast that measures the interaction between probucol and multivitamins is "the difference in differences" from parts (a) and (b):

$(\bar{y}_2 - \bar{y}_1) - (\bar{y}_4 - \bar{y}_3) = 0.36 - 0.14 = 0.22$.

[Note: This is not the only correct answer; reversing the signs in (a) and (b), or in (c), is also correct.]

CHAPTER 12
Linear Regression and Correlation

12.2.1 (a) Graph (iv) shows the strongest negative correlation, followed by (i). The correlation in (ii) is near zero. Graph (iii) shows moderate positive correlation and graph (v) shows the strongest positive correlation. Thus, (iv), (i), (ii), (iii), (v). (The actual correlations are -0.97, -0.63, 0.10, 0.58, and 0.93.)

(b) The correct order is (iv), (iii), (ii).

12.2.2 (b) To compute the sample correlation we must first find the sum of the products of the standardized data. The table below summarizes these intermediate results. The sample correlation, r, is $\frac{1}{(5-1)} \times 1.743 = 0.44$. Note: Using statistical software with the raw data, one will obtain a slightly different answer due to the rounding of the standard deviations listed in the problem.

	x	y	$\frac{(x-\bar{x})}{s_x}$	$\frac{(y-\bar{y})}{s_y}$	$\frac{(x-\bar{x})}{s_x} \times \frac{(y-\bar{y})}{s_y}$
	6	6	1.238	-0.085	-0.105
	1	7	-1.143	0.128	-0.146
	3	3	-0.190	-0.723	0.138
	2	2	-0.667	-0.936	0.624
	5	14	0.762	1.617	1.232
sum	17.0	32.0	0.000	0.000	1.743
mean	3.4	6.4			
sd	2.1	4.7			

12.2.3 The hypotheses are

H_0: There is no correlation between blood urea and uric acid concentration in the population of all healthy persons ($\rho = 0$)

H_A: Blood urea and uric acid concentration are positively correlated ($\rho > 0$)

The test statistic is

$$t_s = r\sqrt{\frac{n-2}{1-r^2}} = 0.2291\sqrt{\frac{282}{1-0.2291^2}} = 3.952.$$

The degrees of freedom are 284 - 2 = 282, so we consult Table 4 using df = 140. We find that $t_{(40, 0.0005)} = 3.361$. Thus, P < 0.0005, so we reject H_0. There is sufficient evidence (P < 0.0005) to conclude that blood urea and uric acid concentration are positively correlated.

12.2.6 (a) The hypotheses are

H_0: $\rho = 0$ (no linear relationship between cob weight and density)

H_A: $\rho \neq 0$ (a linear relationship between cob weight and density)

The test statistic is

$$t_s = r\sqrt{\frac{n-2}{1-r^2}} = -0.9418 \sqrt{\frac{20-2}{1-(-0.9418)^2}} = -11.886$$

Consulting Table 4 with df = 20 - 2 = 18, we find that $t_{0.0005} = 3.922$. Thus, P < 0.001, so we reject H_0 (with $\alpha = 0.05$). There is strong evidence that the population correlation is different from zero; the evidence is that cob weight and plant density are linearly related.

(b) This study was an observational study. The plant densities were merely observed, not manipulated by the researcher.

(c) As an observational study with a statistically significant correlation between cob weight and plant density, we can only say that there is an association between weight and density. We cannot make any causal connection. That is, we do not have evidence that altering the plant density will, in turn, alter cob weight.

12.3.1 (b) $b_1 = r\left(\dfrac{s_Y}{s_X}\right) = 0.993\left(\dfrac{0.637}{21.60}\right) = 0.02928$ ng/min (the rate of incorporation);

$b_0 = 0.83 - (0.02928)(30) \approx -0.05$.

(d) $s_e = \sqrt{\dfrac{SS(resid)}{n-2}} = \sqrt{\dfrac{0.035225}{5}} = 0.0839$.

12.3.3 (c) The slope and intercept of the regression line are

$$b_1 = 0.91074\left(\frac{2.13437}{0.25439}\right) = 7.64124 \approx 7.641;$$

$b_0 = \overline{y} - b_1\overline{x} = 3.05333 - (7.64124)(0.477) = -0.592$.

The fitted regression line is $\hat{Y} = -0.592 + 7.641X$.

$$s_e \approx s_Y\sqrt{1-r^2} = 2.13437\sqrt{1-0.91074^2} = 0.881°C.$$

12.3.6 (a) The slope and intercept of the regression line are

$$b_1 = 0.98139\left(\frac{308.254}{12.095}\right) = 25.011;$$

$$b_0 = 2168.429 - 25.011(62.400) = 607.7.$$

The fitted regression line is $\hat{Y} = 607.7 + 25.01X$.

(b)

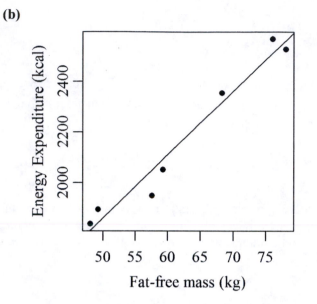

(c) As fat-free mass goes up by 1 kg, energy expenditure goes up by 25.01 Kcal, on average.

(d) $s_e = \sqrt{\dfrac{21026.1}{5}} = 64.85$ kcal.

12.3.10 (b) $r^2 = (0.32725)^2 = 0.107$, so 10.7% of the variation in flow rate is explained by the linear relationship between flow rate and height.

(f) 12 of the 17 residuals (71%) are in the interval (-115.16, 115.16).

12.4.5 We begin by finding the equation of the regression line for the data in Exercise 12.2.7. The slope and intercept of the regression line are

$$b_1 = -0.98754\left(\frac{7.8471}{10.884}\right) = -0.7120;$$

$$b_0 = 23.642 - (-0.7120)(11.500) = 31.83$$

The fitted regression line is $\hat{Y} = 31.83 - 0.712X$.

Substituting X = 15 yields

$$\hat{Y} = 31.83 - (0.7120)(15) = 21.1.$$

Thus, we estimate that the mean fungus growth would be 21.1 mm at a laetisaric acid concentration of 15 µg/ml.

According to the linear model, the standard deviation of fungus growth does not depend on X. Our estimate of this standard deviation, $\sigma_{Y|X}$, is the residual standard deviation from the regression line, s_e.

$$s_e = \sqrt{\frac{SS(resid)}{n-2}} = \sqrt{\frac{16.7812}{10}} = 1.3 \text{ mm.}$$

Thus, we estimate that the standard deviation of fungus growth would be 1.3 mm at a laetisaric acid concentration of 15 µg/ml.

12.4.9 (See Exercise 12.3.10 for b_0 and b_1.)

The estimated mean is

$$\hat{Y} = -153.91 + (4.511)(180) = 658.1 \text{ l/min.}$$

The estimated standard deviation is $s_e = 115.16$ l/min.

12.5.1 (a) We begin by calculating b_1 from the data in Exercise 12.3.1:

$$b_1 = 0.993\left(\frac{0.637}{21.60}\right) = 0.02928 \text{ ng/min}$$

Also, $s_e = \sqrt{\frac{0.035225}{5}} = 0.0839$. The standard error of the slope is $SE_{b_1} = \frac{s_e}{s_x\sqrt{n-1}} = \frac{0.0839}{21.6\sqrt{6}} = 0.00159$.

To construct a 95% confidence interval, we consult Table 4 with $df = n - 2 = 7 - 2 = 5$; the critical value is $t_{5,0.025} = 2.571$. The confidence interval is

$$b_1 \pm t_{0.025} SE_{b_1}$$

$$0.02928 \pm (2.571)(0.00159)$$

$$0.0252 < \beta_1 < 0.0334 \text{ ng/min.}$$

(b) We are 95% confident that the rate at which leucine is incorporated into protein in the population of all *Xenopus* oocytes is between 0.0252 ng/min and 0.0334 ng/min.

12.5.5 The sample slope is $b_1 = 0.98139\left(\dfrac{308.254}{12.095}\right) = 25.011$ and $s_e = \sqrt{\dfrac{21026.1}{5}} = 64.85$. The standard error of the slope is

$$SE_{b_1} = \frac{s_e}{s_x\sqrt{n-1}} = \frac{64.85}{12.095\sqrt{6}} = 2.189.$$

(a) To construct a 95% confidence interval we consult Table 4 with df = $n - 2$ = 7 - 2 = 5; the critical value is $t_{5,0.025} = 2.571$. The confidence interval is

$$b_1 \pm t_{0.025}\, SE_{b_1}$$

$$25.011 \pm (2.571)(2.189)$$

$$19.4 < \beta_1 < 30.6 \text{ kcal/kg.}$$

(b) To construct a 90% confidence interval we consult Table 4 with df = $n - 2 = 5$; the critical value is $t_{5,0.05} = 2.015$. The confidence interval is

$$b_1 \pm t_{0.05}\, SE_{b_1}$$

$$25.011 \pm (2.015)(2.189)$$

$$20.6 < \beta_1 < 29.4 \text{ kcal/kg.}$$

12.5.7 The hypotheses are

H_0: There is no linear relationship between respiration rate and altitude of origin ($\beta_1 = 0$)
H_A: Trees from higher altitudes tend to have higher respiration rates ($\beta_1 > 0$)

The sample slope was found in Exercise 12.3.7 to be

$$b_1 = 0.88665\left(\frac{0.07710}{214.617}\right) = 0.0003185;$$

We note that $b_1 > 0$, so the data do deviate from H_0 in the direction specified by H_A.

From Exercise 12.3.7, the residual standard deviation is

$$s_e = \sqrt{\frac{0.013986}{10}} = 0.0374.$$

The standard error of the slope is

$$SE_{b_1} = \frac{s_e}{s_x\sqrt{n-1}} = \frac{0.0374}{214.617\sqrt{11}} = 0.00005254.$$

The test statistic is

$$t_s = \frac{b_1}{SE_{b_1}} = \frac{0.0003185}{0.00005254} = 6.06.$$

Consulting Table 4 with df $= n - 2 = 12 - 2 = 10$, we find that $t_{0.0005} = 4.587$. Thus, $P < 0.0005$, so we reject H_0. There is sufficient evidence ($P < 0.0005$) to conclude that trees from higher altitudes tend to have higher respiration rates.

12.6.6 Scatterplot (b) shows curvature, so it goes with residual plot (i). In scatterplot (c), the points fan out as X increases, so this scatterplot goes with residual plot (ii). Finally, there are no unusual features in scatterplot (a), which goes with residual plot (iii).

12.7.1 (a) The narrow band (dashed lines) is the confidence band. This band tells us about our uncertainty in the *mean* rectal temperature of cows. The narrowness of the band indicates that we have a great deal of precision in our estimate of the mean rectal temperature given relative humidity; however, individual cows are seen to vary substantially from this line.

(b) The wide band (dotted lines) is the prediction band. 95% of all cows are expected to have rectal temperatures for given humidity values within this band. This band accounts for both the uncertainty in the regression line as well as individual variability among the cows.

(c) With less data both bands would become wider.

12.S.1 Let X = body weight and let Y = ovary weight. We wish to estimate $\sigma_{Y|X=4}$. According to the linear model, $\sigma_{Y|X}$ does not depend on X; thus, our estimate of $\sigma \sigma_{Y|X}$ will be s_e. According to Fact 12.3.1,

$$\frac{s_e}{s_Y} \approx \sqrt{1 - r^2} .$$

We are given that $s_Y = 0.429$ gm and $r = 0.836$. Thus, we have

$$\frac{s_e}{0.429} \approx \sqrt{1 - 0.836^2}$$

Solving this equation yields

$$s_e \approx (0.429)\sqrt{1 - 0.836^2} = 0.24.$$

Hence, we estimate $\sigma_{Y|X=4}$ to be 0.24 gm.

12.S.3 (a) We begin by calculating the fitted regression line for the data in Exercise 12.S.2. The slope and intercept of the regression line are

$$b_1 = -0.8506 \left(\frac{0.31175}{0.11724} \right) = -2.262;$$

$$b_0 = 1.117 - (-2.262)(0.12) = 1.39.$$

The fitted regression line is $\hat{Y} = 1.39 - 2.262X$. Substituting $X = 0.24$ yields

$$\hat{Y} = 1.39 - (2.262)(0.24) = 0.85.$$

Thus, we estimate that the mean yield would be 0.85 kg at a sulfur dioxide concentration of 0.24 ppm.

According to the linear model, $\sigma_{Y|X}$ does not depend on X; thus, our estimate of $\sigma_{Y|X}$ will be s_e. We are given that SS(resid) = 0.2955. The residual standard deviation is

$$s_e = \sqrt{0.2955 / (10)} = 0.17 \text{ kg}$$

Thus, we estimate that the standard deviation of yields would be 0.17 kg at a sulfur dioxide concentration of 0.24 ppm.

12.S.6 (a) $s_e = \sqrt{0.09415 / 5} = 0.137$ cm. The approximate relationship is

$$\frac{s_e}{s_Y} \approx \sqrt{1 - r^2} .$$

Substituting the values from above gives

$$\frac{s_e}{s_Y} = \frac{0.137}{0.21035} = 0.652.$$

From part (a)

$$\sqrt{1 - r^2} = \sqrt{1 - 0.80335^2} = 0.596.$$

We verify that $0.652 \approx 0.596$.

(b) The hypotheses are

H_0: Diameter of forage branch is uncorrelated with wing length ($\rho = 0$)
H_A: Diameter of forage branch is correlated with wing length ($\rho \neq 0$)

The test statistic is

$$t_s = r\sqrt{\frac{n-2}{1-r^2}} = 0.80335\sqrt{\frac{7-2}{1-0.80335^2}} = 3.01.$$

Consulting Table 4 with df = 7 - 2 = 5, we find that $t_{0.02} = 2.757$ and $t_{0.01} = 3.365$. Thus, $0.02 < P < 0.04$, so we reject H_0. There is sufficient evidence $(0.02 < P < 0.04)$ to conclude that there is a positive correlation between diameter of forage branch and wing length (i.e., birds with longer wings tend to prefer larger branches).

UNIT IV

IV.4 $\Pr\{\text{type I error}\} = 1 - \Pr\{\text{no error}\} = 1 - 0.95^3 \approx 1 - 0.857 = 0.143$.

IV.5 (a) $r^2 = 0.23^2 = 0.0529$. We can account for 5.3% of the variability in price by using fiber in a regression model.

(b) $\hat{y} = 17.42 + 0.62 \times 0.63 = 17.42 + 1.63 = 19.05$, so the residual is $17.3 - 19.05 = -1.75$.

(c) The typical size of an error in predicting price is around 3.1 cents/ounce.

IV.7 (a) df between groups $= 2$ so $MS_B = \dfrac{14.889}{2} = 7.445$; $MS_W = \dfrac{185.778}{24} = 7.74$. Thus, the F statistic is $7.445/7.74 = 0.96$.

(b) There is essentially no evidence that the mean number of tree species per 0.1 hectare plot differs along the three rivers.

(c) We need random samples of independent observations from normally distributed populations that have a common standard deviation.

IV.12 (a) $b_1 = -0.9061 \times \left(\dfrac{5.4176}{11.0090}\right) = -0.4459$, $b_0 = 41.6858 - (-0.4459 \times 39.4034) = 59.2557$.
$\hat{y} = 59.256 - 0.446x$

(b) $\hat{y} = 59.256 - 0.446 \times 30 = 45.876$

(c) $\mu_{Y|X=30} = 59.256 - 0.446 \times 30 = 45.876$

(d) Since 30mm falls within the range of observed wing lengths, the prediction is an interpolation.

(e) We would expect hummingbird B's wings to beat about 0.446 Hz less (slower) than hummingbird A's.

CHAPTER 13
A Summary of Inference Methods

13.2.1 The response variable in the study is whether or not a patient shows clinically important improvement; this is a categorical variable. The predictor variable is group membership: clozapine or haloperidol. The two samples are independent and the sample sizes are rather large. Thus, a chi-square test of independence would be appropriate. The null hypothesis of interest is H_0: $p_1 = p_2$, where

p_1 = Pr{clinically important improvement if given clozapine}

and

p_2 = Pr{clinically important improvement if given haloperidol}.

A confidence interval for $p_1 - p_2$ would also be relevant.

13.2.10 A two-sample comparison is called for here. Because of the small sample sizes we would want to be fairly confident that pain relief duration is normally distributed if we were to use a two-sample t-test. The normal probability plots shown below both show some curvature casting doubt on the normality of the population distributions. A Wilcoxon-Mann-Whitney test, or a randomization test, is appropriate.

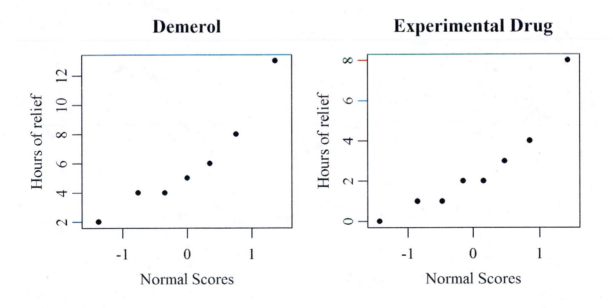

13.2.12 It would be natural to consider correlation and regression with these data. For example, we could regress Y = forearm length on X = height; we could also find the correlation between forearm length and height and test the null hypothesis that the population correlation is zero.